T0322031

Fundamentals of Engineering Plasticity

In this book, Hosford makes the subjects simple by avoiding notations used by specialists in mechanics. R. Hill's authoritative book, *Mathematical Theory of Plasticity* (1950), presented a comprehensive treatment of continuum plasticity theory up to that time; although much of the treatment in this book covers the same ground, it focuses on more recent developments. Hosford has also included recent developments in continuum theory, including a treatment of anisotropy that has resulted from calculations of yielding based on crystallography, analysis of the role of defects, and forming limit diagrams. This text also puts a much greater emphasis on deformation mechanisms and includes chapters on slip and dislocation theory and twinning. This book is useful for those involved in designing sheet metal forming processes. Knowledge of plasticity is essential for the computer simulation of metal forming processes, and understanding the advances in plasticity theory is key to formulating sound analyses.

William F. Hosford is a Professor Emeritus of Materials Science at the University of Michigan. He is the author of numerous research publications and the following textbooks: *Mechanical Behavior of Materials, 2nd Ed.*; *Metal Forming, 4th Ed.* (with Robert M. Caddell); *Materials Science: An Intermediate Text; Reporting Results* (with David C. Van Aken); *Materials for Engineers*; *Solid Mechanics*; *Mechanics of Crystals and Textured Polycrystals*; *Physical Metallurgy, 2nd Ed.*; and *Iron and Steel.* He is also the author of *Wilderness Canoe Tripping.*

FUNDAMENTALS OF ENGINEERING PLASTICITY

William F. Hosford

University of Michigan

CAMBRIDGE
UNIVERSITY PRESS

University Printing House, Cambridge CB2 8BS, United Kingdom

One Liberty Plaza, 20th Floor, New York, NY 10006, USA

477 Williamstown Road, Port Melbourne, VIC 3207, Australia

314-321, 3rd Floor, Plot 3, Splendor Forum, Jasola District Centre, New Delhi - 110025, India

103 Penang Road, #05-06/07, Visioncrest Commercial, Singapore 238467

Cambridge University Press is part of the University of Cambridge.

It furthers the University's mission by disseminating knowledge in the pursuit of education, learning and research at the highest international levels of excellence.

www.cambridge.org
Information on this title: www.cambridge.org/9781107037557

First published 2013

A catalogue record for this publication is available from the British Library

Library of Congress Cataloging in Publication data
Hosford, William F.
Fundamentals of engineering plasticity / William F. Hosford, University of Michigan.
pages cm
Includes bibliographical references and index.
ISBN 978-1-107-03755-7 (hardback)
1. Plasticity. 2. Metal-work. I. Title.
TA418.14.H67 2013
620.1′633–dc23 2012043349

ISBN 978-1-107-03755-7 Hardback

CONTENTS

PREFACE

In 1950, R. Hill wrote an authoritative book, *Mathematical Theory of Plasticity*, that presented a comprehensive treatment of continuum plasticity theory as known at that time. Much of the treatment in this book covers some of the same ground but there is no attempt to treat all the same topics treated by Hill. This book, however, includes more recent developments in continuum theory, including a newer treatment of anisotropy that has resulted from calculations of yielding based on crystallography, analysis of the role of defects, and forming limit diagrams. There is a much greater emphasis on deformation mechanisms, including chapters on slip and dislocation theory and twinning.

This book should provide a useful resource to those involved with designing processes for sheet metal forming. Knowledge of plasticity is essential to those involved in computer simulation of metal forming processes. Knowledge of the advances in plasticity theory are essential in formulating sound analyses.

In writing this book, I have tried to make the subjects simple by avoiding some of the modern notations used by specialists in mechanics.

This book can form the basis for a graduate course in the field of mechanical engineering.

1

AN OVERVIEW OF THE HISTORY OF PLASTICITY THEORY

INTRODUCTION

Plasticity theory deals with yielding of materials, often under complex states of stress. Plastic deformation, unlike elastic deformation, is permanent in the sense that after stresses are removed the shape change remains. Plastic deformation usually occurs almost instantaneously, but creep can be regarded as time-dependent deformation plastic deformation.

There are three approaches to plasticity theory. The approach most widely used is continuum theory. It depends on yield criteria, most of which are simply postulated without regard to how the deformation occurs. Continuum plasticity theory allows predictions of the stress states that cause yielding and the resulting strains. The amount of work hardening under different loading conditions can be compared.

A second approach focuses on the crystallographic mechanisms of slip (and twinning), and uses understanding of these to explain continuum behavior. This approach has been quite successful in predicting anisotropic behavior and how it depends on crystallographic texture. Ever since the 1930s, there has been increasing work bridging the connection between this crystallographic approach and continuum theory.

The third approach to plasticity has been concentrated on how slip and twinning occur. Dislocation theory, first postulated in the 1930s, has given insight and some understanding of how crystalline materials deform by slip. It explains strain hardening, but the connection to continuum theory has been difficult to bridge.

CONTINUUM THEORIES

The theoretical basis for yielding under complex stress states had its origins in the nineteenth century. The first systematic investigation of yielding can be attributed to Tresca [1] who conducted a series of experiments on extrusion and concluded that yielding occurred when the maximum shear stress reached a critical value. He was probably influenced by earlier work of Coulomb [2] on soil mechanics. In 1913, Von Mises [3] proposed his widely used yield criterion. Huber [4] had earlier published essentially the same criterion in Polish, but he may have been writing about fracture and his paper had attracted little attention. Von Mises work was also preceded by Maxwell [5] written in 1856 in an unpublished letter.

In 1937, Nadai [6] showed that the von Mises criterion corresponds to yielding when a critical shear stress is reached on the octahedral planes. It also was shown [7, 8] that the von Mises criterion can be derived, if one assumes that yielding occurs when the elastic distortional energy reaches a critical value. Although this has been taken as proof of the von Mises criterion, there is no fundamental reason for this assumption.

In 1948, Hill [9] proposed the first anisotropic yield criterion. However, it was not until the 1970s that non-quadratic yield criteria [10, 11] were proposed. A non-quadratic modification of Hill's 1948 criterion was proposed in 1979 [12].

CRYSTALLOGRAPHIC BASIS OF PLASTICITY

In 1900, Ewing and Rosenhain [13] showed that plastic deformation occurred by slip. This is the sliding of planes of atoms slide over one another. The planes on which slip occurs are called *slip planes* and the directions of the shear are the *slip directions*. These crystallographic planes and directions are characteristic of a material's crystal structure. The magnitude of the shear displacement is an integral number of inter-atomic distances, so that the lattice is left unaltered. In 1924, Schmid [14] proposed that slip occurs when the shear stress on the slip plane in the slip direction has to reach a critical value. Along with Boas, Schmid published *Kristallplastizität* [15], a classic book on slip.

Calculations of the critical stress to cause slip predicted strengths several orders of magnitude higher than those found experimentally. In 1934, dislocation theory was formulated by three independent scientists to explain this discrepancy [16, 17, 18]. In 1954, Frank and Read [19] showed how slip can generate dislocations. Since the introduction of dislocation theory, it has been realized dislocation climb and cross slip could overcome obstacles and that the intersection of dislocations on different planes is responsible for strain hardening.

In 1938, Taylor [20, 21] developed an upper bound model of the deformation of polycrystals based on the nature of slip. He assumed that every grain must undergo the same shape change. His analysis assumed that the shape change would occur with the minimum amount of slip. In 1951, Bishop and Hill [22, 23] proposed an alternate way of viewing the problem by finding the stress states that are capable of activating enough slip systems to allow every grain to undergo the same shape change. These theories allowed analysis of the deformation of polycrystalline metals.

General Treatments of Plasticity

In 1950, Hill wrote a classic book, *The Mathematical Theory of Plasticity* [24], which covered the basic theory of plasticity and applications to a number of problems. It also introduced a treatment of anisotropic plastic behavior. This was followed by Timoshenko's *History of the*

Strength of Materials in 1953 [25] and Calladine's *Engineering Plasticity* in 1969 [26]. However, since the preceding there have been no new general treatments of plasticity.

NOTE OF INTEREST

Although the book, *Kristallplasttizität*, by Schmid and Boas was first published in 1935, it was available only in German because the Nazis refused to allow it to be translated. Only after World War II, was translation undertaken. In 1950, an English edition was published by Chapman and Hall.

REFERENCES

1. H. Tresca, *Comptes Rendus Acad. Sci. Paris* v. 59 (1864) and v. 64 (1867).
2. C. A. Coloumb, *Mém. Math. et Phys.* v. 7 (1773).
3. R. von Mises, *Göttinger Nachrichten Math-Phys. Klasse* (1913).
4. M. T. Huber, *Czasopismo technische Lemberg* v. 22 (1904).
5. Clerk Maxwell, letter to W. Thompson (1856).
6. A. Nadai, *J. App. Phys.* v 8 (1937).
7. E. Beltrami, *Rend. Inst Lomb.* v. 18 (1885).
8. B. P. Haigh, *Brit. Ass. Reports* Section G (1918).
9. R. Hill, *Proc. Roy. Soc.* v. 193A (1948).
10. R. Hill, *Math. Proc. Camb. Soc.* v. 75 (1979).
11. W. F. Hosford, *J. Appl. Mech. (Trans. ASME ser E.)* v. 39E (1972).
12. W. F. Hosford, *7th North Amer. Metalworking Conf.* SME (1980).
13. J. A. Ewing and W. Rosenhain, *Proc. Roy. Soc* v. A67 (1900).
14. E. Schmid, *Proc. Internat. Cong. Appl. Mech.* Delft (1924).
15. E. Schmid and W. Boas, *Kristallplasttizität*, Springer-Verlag (1935).
16. G. I. Taylor, *Proc. Roy. Soc* v. A145 (1934).
17. M. Polyani, *Z. Physik* v. 89 (1934).
18. E. Orowan, *Z. Physik* v. 89 (1934).
19. F. C. Frank and W. T. Read, *Phys. Rev.* v. 79 (1950).
20. G. I. Taylor, *J. Inst, Metals* v. 62 (1938).
21. G. I. Taylor in *Timoshenko Aniv. Vol.* Macmillan (1938).
22. J. F. W. Bishop and R. Hill, *Phil. Mag Ser.* 7, v. 42 (1951).
23. J. F. W. Bishop and R. Hill, *Phil. Mag Ser.* 7, v. 42 (1951).

24. R. Hill, *The Mathematical Theory of Plasticity*, Oxford University Press (1950).
25. S. P. Timoshenko, *History of Strength of Materials*, McGraw-Hill (1953).
26. C. R. Calladine, *Engineering Plasticity*, Pergamon (1969).

GENERAL REFERENCES

L. M. Kachanov, *Fundamentals of the Theory of Plasticity*, Dover Books (2004).
A.S. Khan and S. Huang, *Continuum Theory of Plasticity*, John Wiley & Sons (1995).
J. Lubliner, *Plasticity Theory*, Macmillan Publishing (1990).
S. Z. Marciniak, J. L. Duncan and S. J. Hu, *Mechanics of Sheet Metal Forming*, Butterworth-Heinemann (2002).
S. Nemat-Nasser, *Plasticity*, Cambridge University Press (2004).
Van Vliet, K. J., *Mechanical Behavior of Materials*, MIT (2006).

2

YIELDING

Of concern in plasticity theory is the *yield strength,* which is the level of stress that causes appreciable plastic deformation. It is tempting to define yielding as occurring at an *elastic limit* (the stress that causes the first plastic deformation) or at a *proportional limit* (the first departure from linearity). However, neither definition is very useful because they both depend on accuracy of strain measurement. The more accurately the strain is measured, the lower is the stress at which plastic deformation and non-linearity can be detected.

To avoid this problem, the onset of plasticity is usually described by an *offset yield strength* that can be measured with more reproducibility. It is found by constructing a straight line parallel to the initial linear portion of the stress strain curve, but offset from it by a strain of $\Delta e = 0.002$ (0.2%). The yield strength is taken as the stress level at which this straight line intersects the stress strain curve (Figure 2.1). The rationale is that if the material had been loaded to this stress and then unloaded, the unloading path would have been along this offset line resulting in a plastic strain of $e = 0.002$ (0.2%). This method of defining yielding is easily reproduced.

If yielding in a tension test is defined by a 0.2% offset, for the purpose of assessing the anisotropy, yielding under any other form of loading must be defined by the plastic strain that involves the same amount of plastic work as the 0.2% offset in tension.

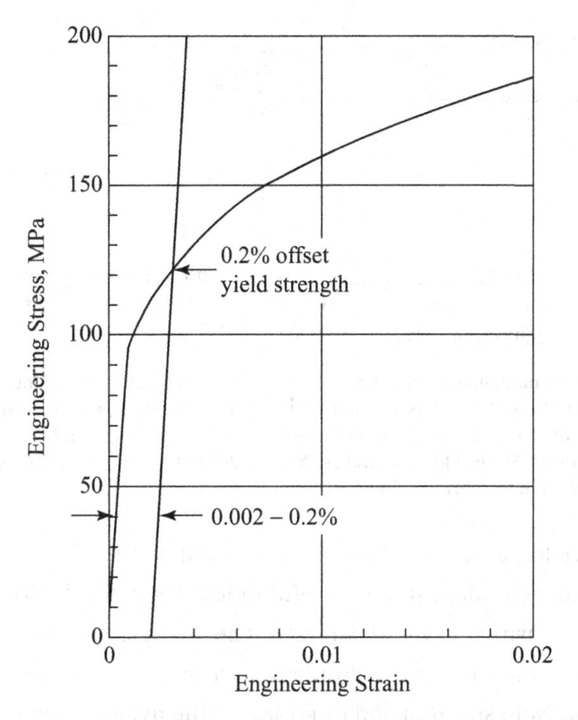

Figure 2.1. The low-strain region of the stress-strain curve for a ductile material. From W. F. Hosford, Mechanical Behavior of Materials, 2nd ed., Cambridge University Press (2010).

Yield points: The stress–strain curves of some materials (for example, low carbon steels and linear polymers), have an initial maximum followed by lower stress as shown in Figures 2.2a and 2.2b. After the initial maximum, at any given instant all of the deformation occurs within a relatively small region of the specimen. For steels, this deforming region is called a *Lüder's* band. Continued elongation occurs by propagation of the Lüder's band along the gauge section, rather than by continued deformation within it. Only after the band has traversed the entire gauge section, does the stress rise again. In the case of linear polymers, the yield strength is usually defined as the initial maximum stress. For steels, the subsequent lower yield strength is used to

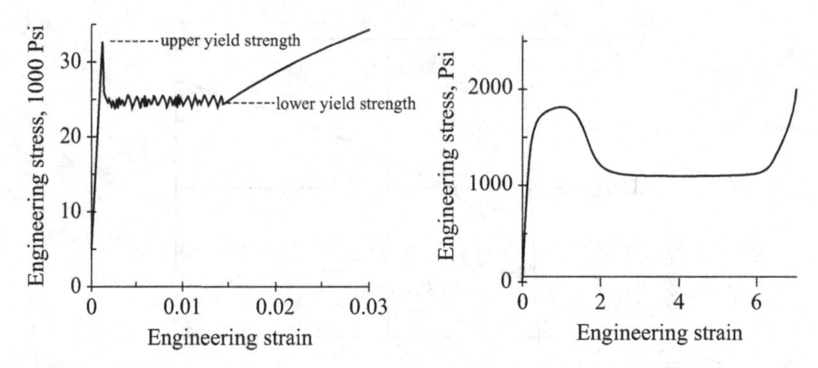

Figure 2.2. Inhomogeneous yielding of low carbon steel (left) and a linear polymer (right). After the initial stress maximum, the deformation in both materials occurs within a narrow band that propagates the length of the gauge section before the stress rises again. From W. F. Hosford and R. M. Caddell, *Metal Forming; Mechanics and Metallurgy*, 4th ed. Cambridge University Press (2007).

describe yielding because the initial maximum stress is too sensitive to specimen alignment to be a useful index. Even so, the lower yield strength is sensitive to the strain rate. The stress level during Lüder's band propagation fluctuates. Some laboratories report the minimum level as the yield strength and other use as the average level.

IDEALIZATION OF YIELDING BEHAVIOR

Typical tensile load-extension behavior with unloading and reloading is shown schematically in Figure 2.3a. Idealization of this behavior

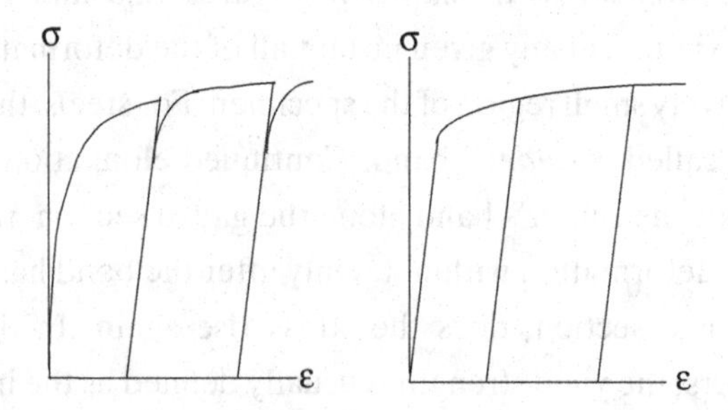

Figure 2.3. Idealization of yielding. Actual loading and unloading stress strain curves (A) are often idealized (B) by assuming sharp yielding on reloading after unloading.

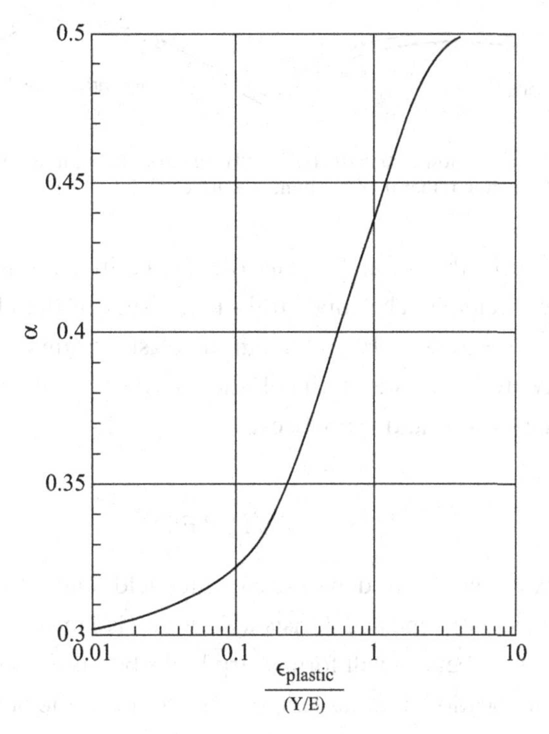

Figure 2.4 Change in the stress ratio, $\alpha = \sigma_y/\sigma_x$, for plane strain, $\varepsilon_y = 0$, as a function of strain. From W. F. Hosford and R. M. Caddell, *Metal Forming; Mechanics and Metallurgy*, 4th ed. Cambridge University Press (2007).

(Figure 2.3b) has a sharp initial yield stress (a) and assumes sharp yielding on reloading after unloading.

ELASTIC-PLASTIC TRANSITION

The transition from elastic to plastic flow is gradual as illustrated in Figure 2.4 for plane-strain deformation with $\varepsilon_y = 0$ and $\sigma_z = 0$. For elastic deformation, $\alpha = \nu$ and for fully plastic deformation $\alpha = 0.5$. In this figure, the ε_x is normalized by the ratio of the yield strength to the modulus. Note that 95% of the change from elastic to plastic deformation occurs when the plastic strain is three times the elastic strain.

Figure 2.5 Bending of a low carbon steel wire will result in kinks because of the tendency to localize deformation rather than a continuous curve as with copper wire.

For a material that strain hardens, there is additional elastic deformation after yielding. The total strain is the sum of the elastic and plastic parts, $e = e_e + e_p$. Even though the elastic strain may be very small relative to the plastic strain, elastic recovery on unloading controls residual stresses and springback.

NOTE OF INTEREST

A simple experiment that demonstrates the yield point effect can be made with pieces of annealed florists wire, which is a low-carbon steel. When the wire is bent, it will form sharp kinks because once yielding occurs at one location, it takes less force to continue the bend at that location than to initiate bending somewhere else. On the other hand, copper wire that has no yield point will bend in a continuous arc (see Figure 2.5).

REFERENCES

1. W. F. Hosford, *Mechanical Behavior of Materials*, 2nd ed., Cambridge University Press (2010).
2. W. F. Hosford and R. M. Caddell, *Metal Forming; Mechanics and Metallurgy*, 4th ed., Cambridge University Press (2007).

3

STRESS AND STRAIN

An understanding of stress and strain is essential for the analysis of plastic deformation. Often the words *stress* and *strain* are used synonymously by the non-scientific public. In engineering usage, however, stress is the intensity of force and strain is a measure of the amount of deformation.

STRESS

Stress is defined as the intensity of force at a point.

$$\sigma = \partial F/\partial A \quad \text{as } \partial A \to 0. \tag{3.1}$$

If the stress is the same everywhere in a body,

$$\sigma = F/A. \tag{3.2}$$

As shown in Figure 3.1, there are nine components of stress. A normal stress component is one in which the force is acting normal to the plane. It may be tensile or compressive. A shear stress component is one in which the force acts parallel to the plane.

Stress components are defined with two subscripts. The first denotes the normal to the plane on which the force acts and the second is the direction of the force.[1] For example, σ_{xx} is a tensile stress on

[1] The use of the opposite convention should cause no problem because $\sigma_{ij} = \sigma_{ji}$.

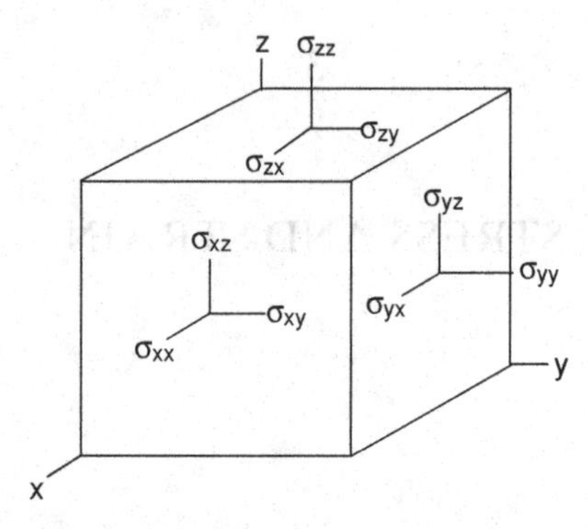

Figure 3.1. Nine components of stress acting on an infinitesimal element. From W. F. Hosford and R. M. Caddell, *Metal Forming: Mechanics and Metallurgy,* 4th ed. Cambridge University Press, 2011.

the place normal to x, in the x-direction. A shear stress acting on the x-plane in the y-direction is denoted σ_{xy}.

Repeated subscripts, (for example, σ_{xx}, σ_{yy}, σ_{zz}) indicate normal stresses. They are tensile if both subscripts are positive or both are negative. If one is positive and the other is negative, they are compressive. Mixed subscripts (for example, σ_{zx}, σ_{xy}, σ_{yz}) denote shear stresses. A state of stress in tensor notation is expressed as

$$\sigma_{ij} = \begin{vmatrix} \sigma_{xx} & \sigma_{yx} & \sigma_{zx} \\ \sigma_{xy} & \sigma_{yy} & \sigma_{zx} \\ \sigma_{xz} & \sigma_{yz} & \sigma_{zz} \end{vmatrix}, \quad 3.3$$

where *i* and *j* are iterated over x, y and z. Except where tensor notation is required, it is simpler to use a single subscript for a normal stress and denote a shear stress by τ. For example, $\sigma_x \equiv \sigma_{xx}$ and $\tau_{xy} \equiv \sigma_{xy}$.

Stress Transformation

Stress components expressed along one set of orthogonal axes may be expressed along any other set of axes. Consider resolving the stress component, $\sigma_y = F_y/A_y$, onto the x′ and y′ axes as shown in Figure 3.2.

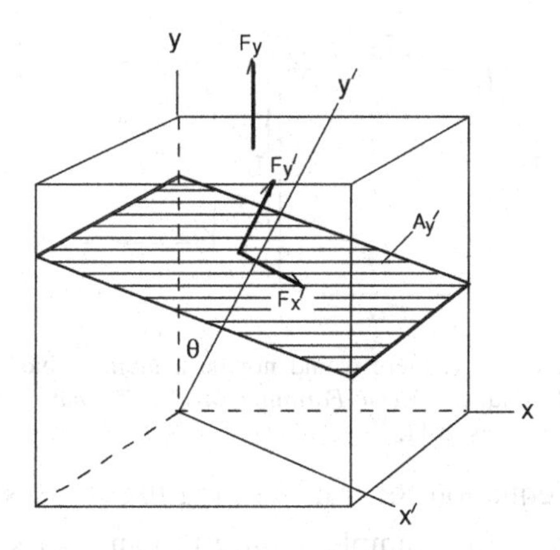

Figure 3.2. The stresses acting on a plane, A', under a normal stress, σ_y. From W. F. Hosford and R. M. Caddell, *Metal Forming: Mechanics and Metallurgy,* 4th ed. Cambridge University Press, 2011.

The force, $F_{y'}$, acts in the y′ direction is $F_y' = F_y \cos\theta$ and the area normal to y′ is

$$A_y' = A_y/\cos\theta, \text{ so}$$

$$\sigma_{y'} = F_{y'}/A_{y'} = F_y \cos\theta/(A_y/\cos\theta) = \sigma_y \cos^2\theta. \qquad 3.4a$$

Similarly,

$$\tau_{y'x'} = F_{x'}/A_{y'} = F_y \sin\theta/(A_y/\cos\theta) = \sigma_y \cos\theta \sin\theta. \qquad 3.4b$$

Note that transformation of stresses requires two sine and/or cosine terms.

Pairs of shear stresses with the same subscripts in reverse order are always equal (for example, $\tau_{ij} = \tau_{ji}$). This is illustrated in Figure 3.3 by a simple moment balance on an infinitesimal element. Unless $\tau_{ij} = \tau_{ji}$, there would be an infinite rotational acceleration. Therefore,

$$\tau_{ij} = \tau_{ji}. \qquad 3.5$$

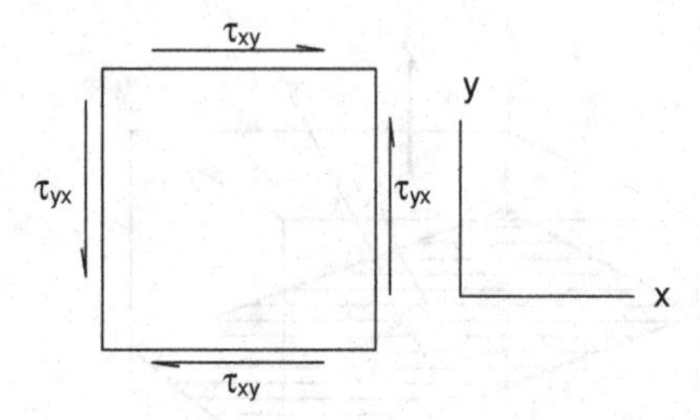

Figure 3.3. Unless $\tau_{xy} = \tau_{yx}$, there would not be a moment balance. From W. F. Hosford and R. M. Caddell, *Metal Forming: Mechanics and Metallurgy,* 4th ed. Cambridge University Press, 2011.

The general equation for transforming the stresses from one set of orthogonal axes (for example, *n. m, p*) to another set of axes (for example, *i, j, k*), is

$$\sigma_{ij} = \sum_{n=1}^{3}\sum_{m=1}^{3} \ell_{im}\ell_{jn}\sigma_{mn}. \qquad 3.6$$

Here, the term ℓ_{im} is the cosine of the angle between the i and the m axes and the term ℓ_{jn} is the cosine of the angle between the j and n axes. This is often written more simply as

$$\sigma_{ij} = \ell_{in}\ell_{jn}\sigma_{mn}, \qquad 3.7$$

with the summation implied. Consider transforming stresses from the x, y, z axis system to the x′, y′, z′ system shown in Figure 3.4.

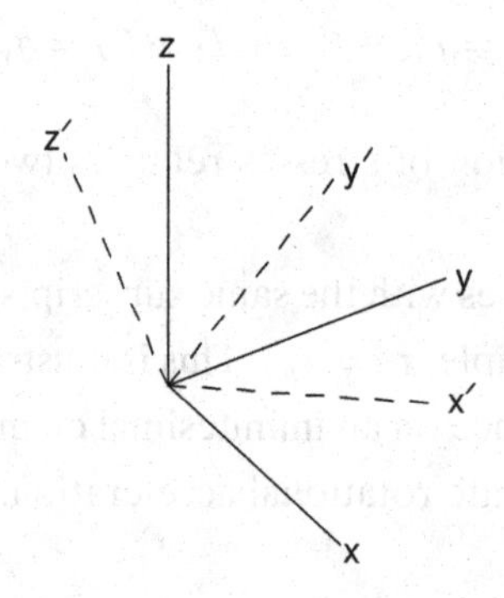

Figure 3.4. Two orthogonal coordinate systems. From W. F. Hosford, *Mechanical Behavior of Materials,* 2nd ed., Cambridge University Press, 2010.

Using equation 3.7,

$$\sigma_{x'x'} = \ell_{x'x}\ell_{x'x}\sigma_{xx} + \ell_{x'x}\ell_{x'y}\sigma_{xy} + \ell_{x'y}\ell_{x'x}\sigma_{yx} + \ell_{x'y}\ell_{x'y}\sigma_{yy} + \ell_{x'x}\ell_{x'z}\sigma_{xz} + \ell_{x'z}\ell_{x'x}\sigma_{zx} + \ell_{x'z}\ell_{x'z}\sigma_{zz} \qquad 3.8a$$

and

$$\sigma_{x'y'} = \ell_{x'x}\ell_{y'x}\sigma_{xx} + \ell_{x'y}\ell_{y'x}\sigma_{xy} + \ell_{x'x}\ell_{x'y}\sigma_{yx} + \ell_{x'y}\ell_{y'y}\sigma_{yy} + \ell_{x'z}\ell_{y'z}\sigma_{yz} + \ell_{x'x}\ell_{y'z}\sigma_{xz} + \ell_{x'y}\ell_{y'z}\sigma_{zy} + \ell_{x'z}\ell_{y'z}\sigma_{zz} \qquad 3.8b$$

These can be simplified to

$$\sigma_{x'} = \ell_{x'x}^2\sigma_x + \ell_{x'y}^2\sigma_y + \ell_{x'z}^2\sigma_z + 2\ell_{x'y}\ell_{x'z}\tau_{yz} + 2\ell_{x'z}\ell_{x'x}\tau_{zx} + 2\ell_{x'x}\ell_{x'y}\tau_{xy} \qquad 3.9a$$

and

$$\tau_{x'y'} = \ell_{x'x}\ell_{y'x}\sigma_x + \ell_{x'y}\ell_{y'y}\sigma_y + \ell_{x'z}\ell_{y'z}\sigma_z + (\ell_{x'y}\ell_{y'z} + \ell_{x'z}\ell_{y'y})\tau_{yz} + (\ell_{x'z}\ell_{y'x} + \ell_{x'x}\ell_{y'z})\tau_{zx} + (\ell_{x'x}\ell_{y'y} + \ell_{x'y}\ell_{y'x})\tau_{xy} \qquad 3.9b$$

It is always possible to find a set of axes along which the shear stress terms vanish. In this case, the normal stresses, σ_1, σ_2 and σ_3, are called principal stresses. They are the roots of

$$\sigma_P^3 - I_1\sigma_P^2 - I_2\sigma_p - I_3 = 0 \qquad 3.10$$

where I_1, I_2, and I_3 are called the *invariants* of the stress tensor. They are

$$I_1 = \sigma_{xx} + \sigma_{yy} + \sigma_{zz}, \qquad 3.11a$$

$$I_2 = \sigma_{yz}^2 + \sigma_{zx}^2 + \sigma_{xy}^2 - \sigma_{yy}\sigma_{zz} - \sigma_{zz}\sigma_{xx} - \sigma_{xx}\sigma_{yy} \text{ and} \qquad 3.11b$$

$$I_3 = \sigma_{xx}\sigma_{yy}\sigma_{zz} + 2\sigma_{yz}\sigma_{zx}\sigma_{xy} - \sigma_{xx}\sigma_{yz}^2 + \sigma_{yy}\sigma_{zx}^2 + \sigma_{zz}\sigma_{xx}^2. \qquad 3.11c$$

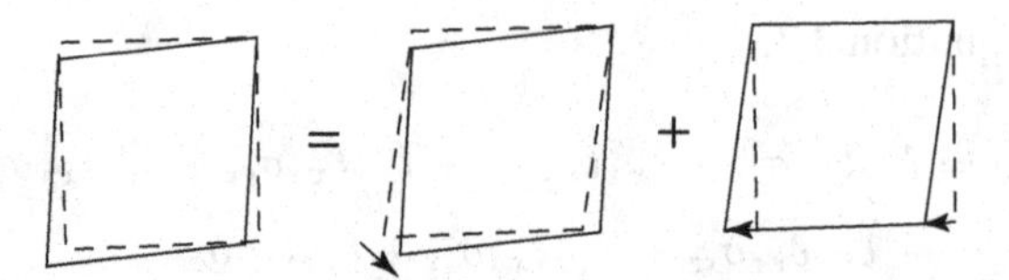

Figure 3.5. Illustration of shear and rotation. From W. F. Hosford, *Mechanical Behavior of Materials,* 2nd ed. Cambridge University Press, 2010.

SMALL STRAINS

When a body deforms, it often also undergoes translation and rotation as well. Strains must be defined in such a way as to exclude these effects (Figure 3.5). Figure 3.6 shows a small two-dimensional element, ABCD, deformed into A′B′C′D′ where the displacements are u and v. The normal strain, e_{xx}, is defined as

$$e_{xx} = (A'C' - AC)/AC = A'C'/AC - 1. \quad 3.12$$

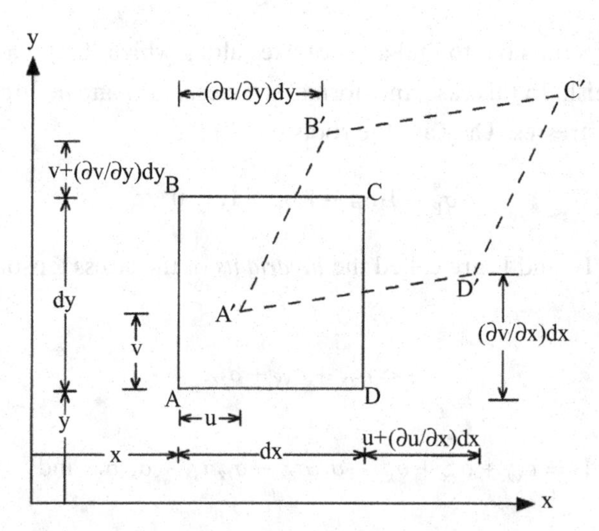

Figure 3.6. Distortion of a two-dimensional element. From W. F. Hosford and R. M. Caddell, *Metal Forming: Mechanics and Metallurgy,* 4th ed. Cambridge University Press, 2011.

Neglecting the rotation,

$$e_{xx} = A'C'/AC - 1 = \frac{dx - u + u + (\partial u/\partial x)dx}{dx} - 1 \text{ or}$$

$$e_{xx} = \partial u/\partial x. \tag{3.13}$$

Similarly, $e_{yy} = \partial v/\partial y$ and $e_{zz} = \partial w/\partial z$ for a three-dimensional case.

The shear strain are associated with the angles between AD and A′D′ and between AB and A′B′. For small deformations,

$$\angle^{AD}_{A'D'} \approx \partial v/\partial x \quad \text{and} \quad \angle^{AB}_{A'B'} = \partial u/\partial y. \tag{3.14}$$

The total shear strain is the sum of these two angles,

$$\gamma_{xy} = \gamma_{yx} = \frac{\partial u}{\partial y} + \frac{\partial v}{\partial x}. \tag{3.15a}$$

Similarly,

$$\gamma_{yz} = \gamma_{zy} = \frac{\partial v}{\partial z} + \frac{\partial w}{\partial y} \text{ and} \tag{3.15b}$$

$$\gamma_{zx} = \gamma_{xz} = \frac{\partial w}{\partial x} + \frac{\partial u}{\partial z}. \tag{3.15c}$$

This definition of shear strain, γ, is equivalent to the simple shear measured in a torsion of shear test.

If the displacements, $\partial u/\partial y$ and $\partial v/\partial x$, are not equal, there is a rotation about the z-axis,

$$\omega_z = (1/2)(\partial u/\partial y - \partial v/\partial x) \tag{3.16}$$

where ω_z is the clockwise rotation in radians. Similar rotations about the x and y axes are

$$\omega_x = (1/2)(\partial v/\partial z - \partial w/\partial y) \quad \text{and} \quad \omega_z = (1/2)(\partial w/\partial x - \partial u/\partial z). \tag{3.17}$$

Such rotations result in the formation of crystallographic textures. The preceding treatment of strains is limited to cases where the strains are small enough that rotation of the axes relative to each other may be neglected. Chin et al. [1] developed an analysis for the large strains that may result from slip.

STRAIN TENSOR

If tensor shear strains, ε_{ij}, are defined as

$$\varepsilon_{ij}, = (1/2)\gamma_{ij}, \tag{3.18}$$

small shear strains form a tensor,

$$\varepsilon_{ij} = \begin{vmatrix} \varepsilon_{xx} & \varepsilon_{yx} & \varepsilon_{zx} \\ \varepsilon_{xy} & \varepsilon_{yy} & \varepsilon_{zy} \\ \varepsilon_{xz} & \varepsilon_{yz} & \varepsilon_{zz} \end{vmatrix}. \tag{3.19}$$

Because small strains form a tensor, they can be transformed from one set of axes to another in a way identical to the transformation of stresses. Mohr's circle relations can be used. It must be remembered, however, that $\varepsilon_{ij} = \gamma_{ij}/2$ and that the transformations hold only for small strains. If $\gamma_{yz} = \gamma_{zx} = 0$,

$$\varepsilon_{x'} = \varepsilon_x \ell_{x'x}^2 + \varepsilon_y \ell_{x'y}^2 + \gamma_{xy} \ell_{x'x} \ell_{x'y} \tag{3.20}$$

$$\text{and } \gamma_{x'y'} = 2\varepsilon_x \ell_{x'x} \ell_{y'x} + 2\varepsilon_y \ell_{x'y} \ell_{y'y} + \gamma_{xy}(\ell_{x'x} \ell_{y'y} + \ell_{y'x} \ell_{x'y}). \tag{3.21}$$

The principal strains can be found from the Mohr's circle equations for strains,

$$\varepsilon_{1,2} = \frac{\varepsilon_x + \varepsilon_y}{2} \pm (1/2)\left[(\varepsilon_x - \varepsilon_y)^2 + \gamma_{xy}^2\right]^{1/2}. \tag{3.22}$$

Strains on other planes are given by

$$\varepsilon_{x,y} = (1/2)(\varepsilon_1 + \varepsilon_2) \pm (1/2)(\varepsilon_1 - \varepsilon_2)\cos 2\theta \tag{3.23}$$

$$\text{and } \gamma_{xy} = (\varepsilon_1 - \varepsilon_2)\sin 2\theta. \tag{3.24}$$

NOTE OF INTEREST

The concepts of tensors is usually attributed to the work of a German mathemetician, Karl Friedrich Gauss (1777–1855). The word "tensor" itself was introduced in 1846 by William Roland to describe something different from what is now meant by a tensor. In 1898, the contemporary usage was brought in by Woldemar Voight.

The simplification of the notation for tensor transformation of $\sigma_{ij} = \sum_{n=1}^{3} \sum_{m=1}^{3} \ell_{im}\ell_{jn}\sigma_{mn}$ to $\sigma_{ij} = \ell_{im}\,\ell_{jn}\sigma_{mn}$ has been attributed to Albert Einstein.

REFERENCE

1. G. Y. Chin, R. N. Thurston, and E. A. Nesbitt, *TMS-AIME* v 236 (1966).

4

ISOTROPIC YIELD CRITERIA

INTRODUCTION

Plasticity theory deals with yielding of materials under complex stress states. It allows prediction of stress states that cause yielding and it predicts the shape change that accompanies yielding. It also allows use of tensile test data to predict the strain hardening that occurs when a material is deformed under any other stress state.

ISOTROPIC YIELD CRITERIA

A *yield criterion* is a mathematical description of the stresses under which yielding occurs. The most general form of a yield criterion is

$$f(\sigma_x, \sigma_y, \sigma_z, \tau_{yz}, \tau_{zx}, \tau_{xy}) = C, \qquad 4.1$$

where C is a material constant. For an isotropic material, this can be expressed in terms of principal stresses,

$$f(\sigma_1, \sigma_2, \sigma_3) = C. \qquad 4.2$$

Yielding of most solids is independent of the sign of the stress state. Reversing the signs of all the stresses has no effect on yielding. This is consistent with the observation that, for most materials, the yield strengths in tension and compression are equal. This may not be true

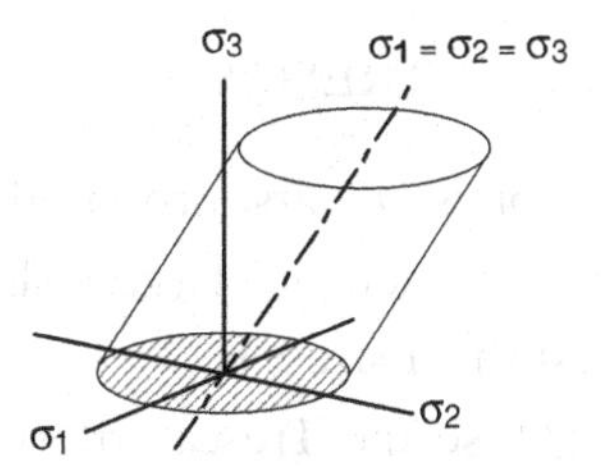

Figure 4.1. A yield locus is the surface of a body in three-dimensional stress space. Stress states on the locus cause yielding. Those inside of the locus will not cause yielding. From W. F. Hosford, *The Mechanics of Crystals and Textured Polycrystals*, Oxford University Press (1993).

when the loading path changes during deformation. Directional difference in yielding behavior after prior staining is called the *Bauschinger effect*. It is also not true if mechanical twinning is an important deformation mechanism (see Chapter 11).

For most solid materials, it is reasonable to assume that yielding is independent of the level of mean normal stress, σ_m,

$$\sigma_m = (\sigma_1 + \sigma_2 + \sigma_3)/3. \quad 4.3$$

Chapter 14 deals with pressure-dependent yielding. Assuming that yielding is independent of σ_m is equivalent to assuming that plastic deformation causes no volume change. This assumption of constancy of volume is certainly reasonable for crystalline materials that deform by slip and twinning because these mechanisms involve only shear. With this simplification, the yield criteria for isotropic materials must be of the form

$$f[(\sigma_2 - \sigma_3), (\sigma_3 - \sigma_1), (\sigma_1 - \sigma_2)] = C. \quad 4.4$$

In terms of the Mohr's stress circle diagrams, only the sizes of the Mohr's circles (rather than their positions) are of importance in determining whether yielding will occur. In three-dimensional stress space (σ_1 vs. σ_2 vs. σ_3), the locus can be represented by a cylinder parallel to the line $\sigma_1 = \sigma_2 = \sigma_3$ as shown in Figure 4.1.

TRESCA

The simplest yield criterion is one first proposed by Tresca [1], which states that yielding occurs when the largest shear stress reaches a critical value. The largest shear stress is

$\tau_{max} = (\sigma_{max} - \sigma_{min})/2$, so the Tresca criterion can be expressed as

$$\sigma_{max} - \sigma_{min} = C. \qquad 4.5$$

With the convention that $\sigma_1 \geq \sigma_2 \geq \sigma_3$, this can be written as

$$\sigma_1 - \sigma_3 = C. \qquad 4.6$$

The constant, C, can be found by considering a tension test in which a tension test. At yielding $\sigma_1 = Y$, where Y is the yield strength and $\sigma_2 = \sigma_3 = 0$. Substituting $\sigma_3 = 0$ and $C = Y$ into equation 3.6, the Tresca criterion may be expressed as

$$\sigma_1 - \sigma_3 = Y. \qquad 4.7$$

For pure shear, $\sigma_1 = -\sigma_3 = k$, where k is the shear yield strength in shear. Substituting $k = Y/2$ into equation 3.7,

$$\sigma_1 - \sigma_3 = 2k = C. \qquad 4.8$$

Letting σ_y and σ_x be principal stresses, the combinations of σ_y and σ_x that will cause yielding with $\sigma_z = 0$ can be plotted. To do this, the σ_y vs. σ_x stress space can be divided into six sectors as shown in Figure 4.2. In the six sectors, the following conditions are appropriate:

I $\sigma_x > \sigma_y > \sigma_z = 0$, so $\sigma_1 = \sigma_x$, $\sigma_3 = \sigma_z = 0$, so $\sigma_x = Y$
II $\sigma_y > \sigma_x > \sigma_z = 0$, so $\sigma_1 = \sigma_y$, $\sigma_3 = \sigma_z = 0$, so $\sigma_y = Y$
III $\sigma_y > \sigma_z = 0 > \sigma_x$, so $\sigma_1 = \sigma_y$, $\sigma_3 = \sigma_x$, so $\sigma_y - \sigma_x = Y$
IV $\sigma_z = 0 > \sigma_y > \sigma_x$, so $\sigma_1 = 0$, $\sigma_3 = \sigma_y$, so $0 - \sigma_x = Y$
V $\sigma_z = 0 > \sigma_x > \sigma_y$, so $\sigma_1 = 0$, $\sigma_3 = \sigma_x$, so $0 - \sigma_y = Y$
VI $\sigma_x > \sigma_z = 0 > \sigma_y$, so $\sigma_1 = \sigma_x$, $\sigma_3 = \sigma_y$, so $\sigma_x - \sigma_y = Y$

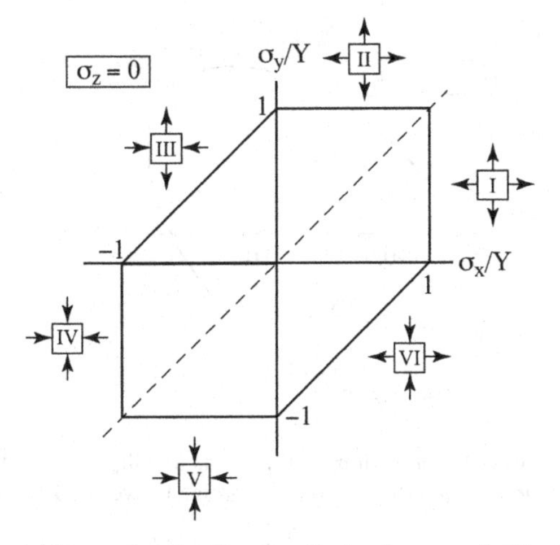

Figure 4.2. The yield locus for the Tresca criterion for $\sigma_z = 0$. The Tresca criterion predicts that the intermediate principal stress has no effect on yielding. For example, in sector I, the value of σ_y has no effect on the value of σ_x required for yielding. Only if σ_y is negative or if it is higher than σ_x does it has an influence. From W. F. Hosford, *Mechanical Behavior of Materials*, 2nd ed., Cambridge University Press (2010).

VON MISES CRITERION

It might seem reasonable to assume that yielding would be affected by the intermediate principal stress. Yielding cannot depend on the average of the diameters of the three Mohr's circles, $[(\sigma_1 - \sigma_2) + (\sigma_2 - \sigma_3) + (\sigma_1 - \sigma_3)]/3$, because the intermediate stress term, σ_2, drops out of the average, $[(\sigma_1 - \sigma_2) + (\sigma_2 - \sigma_3) + (\sigma_1 - \sigma_3)]/3 = (2/3)(\sigma_1 - \sigma_3)$, so an average diameter criterion reduces to the Tresca criterion. The effect of the intermediate principal stress can, however, be included by assuming that yielding depends on the root-mean-square diameter of the three Mohr's circles. This is the von Mises criterion [2] and can be expressed as

$$\{[(\sigma_2 - \sigma_3)^2 + (\sigma_3 - \sigma_1)^2 + (\sigma_1 - \sigma_2)^2]/3\}^{1/2} = C. \qquad 4.9$$

Note that each term is squared so that the convention, $\sigma_1 \geq \sigma_2 \geq \sigma_3$, is not necessary. Again, the material constant, C, can be evaluated by considering a uniaxial tension test. At yielding, $\sigma_1 = Y$ and

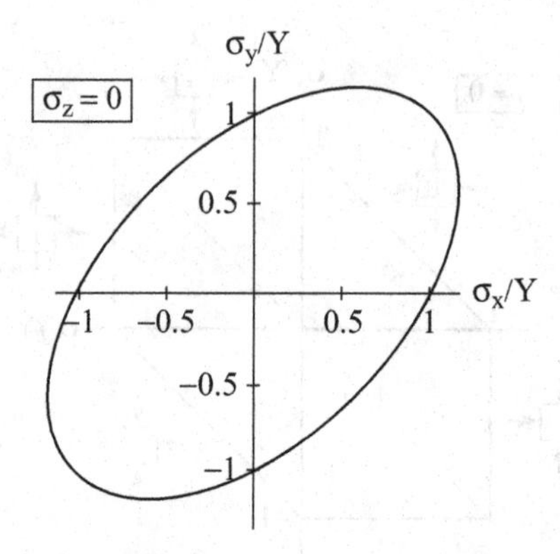

Figure 4.3. The Mises criterion with $\sigma_z = 0$ plots as an ellipse. From W. F. Hosford, *Mechanical Behavior of Materials*, 2nd ed., Cambridge University Press (2010).

$\sigma_2 = \sigma_3 = 0$. Substituting $[0^2 + (-Y)^2 + Y^2]/3 = C^2$ or $C = (2/3)^{1/3}Y$, equation 4.9 becomes

$$(\sigma_2 - \sigma_3)^2 + (\sigma_3 - \sigma_1)^2 + (\sigma_1 - \sigma_2)^2 = 2Y^2. \quad 4.10$$

For a state of pure shear, $\sigma_1 = -\sigma_3 = k$ and $\sigma_2 = 0$. Substituting into equation 4.10,

$$(-k)^2 + [(-k) - k]^2 + k^2 = 2Y^2, \text{ so}$$

$$k = Y/\sqrt{3}. \quad 4.11$$

Equation 4.10 can be simplified when one of the principal stresses is zero (plane-stress conditions). Substituting $\sigma_3 = 0$, $\sigma_1^2 + \sigma_2^2 - \sigma_1\sigma_2 = Y^2$, which plots as an ellipse (Figure 4.3). With further substitution of $\alpha = \sigma_2/\sigma_1$,

$$\sigma_1 = Y/(1 - \alpha + \alpha^2)^{1/2}. \quad 4.12$$

The largest possible ratio of σ_1/Y at yielding with plane stress ($\sigma_3 = 0$), corresponds to the minimum value $1 - \alpha + \alpha^2$ in equation 4.12. Differentiating and setting to zero, $d(1 - \alpha + \alpha^2)/d\alpha = -1 + 2\alpha = 0$, so $\alpha = -1/2$. Substituting into equation 4.12, the maximum value of

$$\sigma_1/Y \text{ is } \sigma_1/Y = [1 - 1/2 + (1/2)^2]^{-1/2} = \sqrt{(4/3)} = 1.155.$$

The von Mises yield criterion can also be expressed in terms of stresses that are not principal stresses. In this case, it includes shear terms.

$$(\sigma_y - \sigma_z)^2 + (\sigma_z - \sigma_x)^2 + (\sigma_x - \sigma_y)^2 + 6\left(\tau_{yz}^2 + \tau_{zx}^2 + \tau_{xy}^2\right) = 2Y^2, \quad 4.13$$

where x, y, and z are not principal stress axes.

EXPERIMENTS

One of the great difficulties in experimentally determining yield criteria is that very few metals are isotropic. Annealing does not remove crystallographic textures. Lode [3] tested thin wall tubes under combined tension and internal pressure. Taylor and Quinney [4] tested thin walled tubes under combined torsion and compression. Both sets of experimental results lie between the the Tresca and von Mises criteria.

OTHER ISOTROPIC YIELD CRITERIA

The Mises and Tresca criteria are not the only possible isotropic yield criteria. Theoretical analysis based on a crystallographic model as well as the experimental data lie between the two and can be represented [5, 6] by

$$|\sigma_2 - \sigma_3|^a + |\sigma_3 - \sigma_1|^a + |\sigma_1 - \sigma_2|^a = 2Y^a. \quad 4.14$$

For $a = 2$ and $a = 4$, this criterion reduces to von Mises. It reduces to Tresca for $a = 1$ and as $a \to \infty$. For exponents between 1 and 2, and exponents greater than 4, this criterion predicts yield loci between Tresca and Mises. Figure 4.4 shows the predictions of this criterion for low values of a. The loci corresponding to exponents between 1 and 2 are duplicated by exponents between 4 and ∞. Exponents between 2 and 4 extend the yield locus further into the first quadrant. The furthest extent occurs with $a = 2.767$.

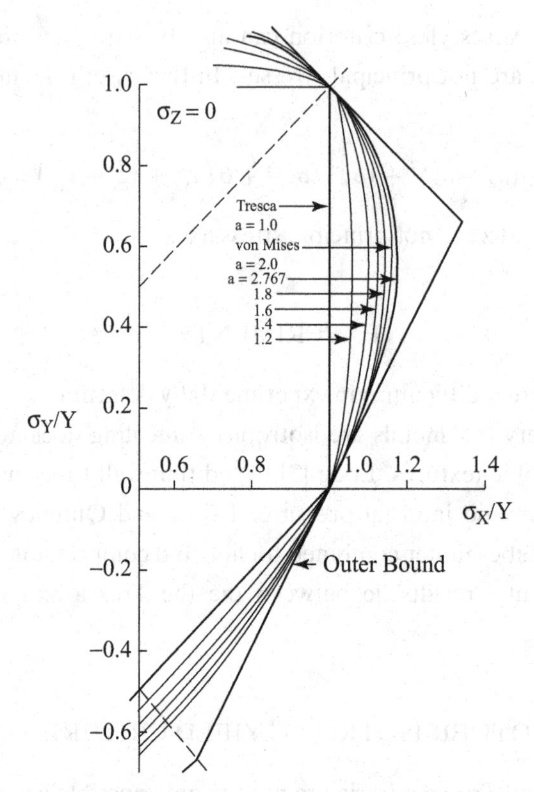

Figure 4.4. Yield loci predicted by equation 4.14. From W. F. Hosford in *Controlling Sheet Metal Forming Processes, Proc. 15th Biennial Congress of the IDDRG*, ASM International (1988).

Figure 4.5 summarizes the effect of the exponent, a, on the extension of the yield locus into the first quadrant. For every a-value between 1 and 2.767, there is an a- value over 2.767 that extends the locus the same amount.

Figure 4.6 shows the effects of an exponent larger than 2. If the exponent, a, is an even integer, equation 3.14 can be written without the absolute magnitude signs as

$$(\sigma_2 - \sigma_3)^{\mathrm{a}} + (\sigma_3 - \sigma_1)^{\mathrm{a}} + (\sigma_1 - \sigma_2)^{\mathrm{a}} = 2Y^{\mathrm{a}}. \qquad 4.15$$

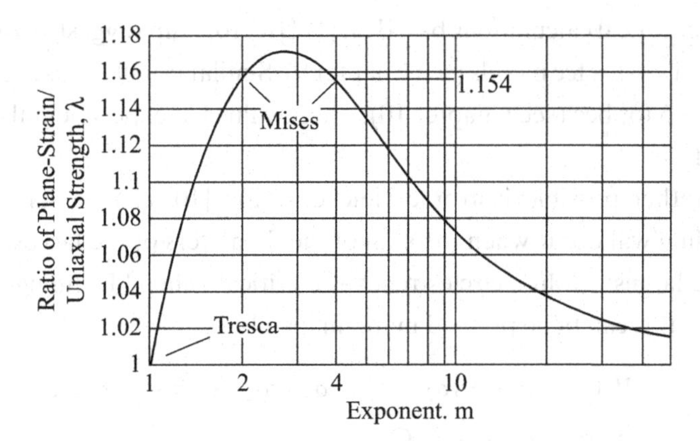

Figure 4.5. The ratio of yield strengths in plane strain tension and uniaxial tension. From W. F. Hosford in *Controlling Sheet Metal Forming Processes, Proc. 15th Biennial Congress of the IDDRG*, ASM International (1988).

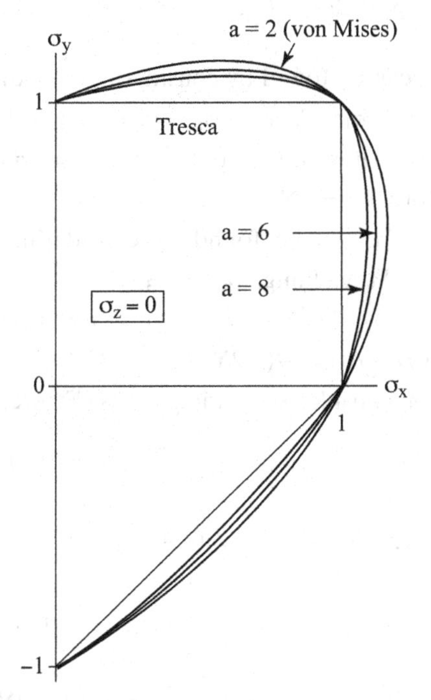

Figure 4.6. Yield loci for $(\sigma_2 - \sigma_3)^a + (\sigma_3 - \sigma_1)^a + (\sigma_1 - \sigma_2)^a = 2Y^a$ with several values of a. Note that the von Mises criterion corresponds to a = 2 and the Tresca criterion to a = 1. From W. F. Hosford, *The Mechanics of Crystals and Textured Polycrystals*, Oxford University Press (1993).

Theoretical calculations based on {111}<110> slip suggest an exponent of a = 8 for fcc metals (see Chapter 9). Similar calculations suggest that a = 6 for bcc (see Chapter 10). These values fit experimental data as well.

Another possible isotropic yield criterion [10] can be stated as "Yielding will occur when the sum of the diameters of the largest and second largest Mohr's circles reaches a critical value." Defining $\sigma_1 \geq \sigma_2 \geq \sigma_3$, this can be expressed mathematically as:

$$1)\ \text{If } (\sigma_1 - \sigma_2) \geq (\sigma_2 - \sigma_3),\ (\sigma_1 - \sigma_2) + (\sigma - \sigma_3) = C$$

$$\text{or } 2\sigma_1 - \sigma_2 - \sigma_3 = C_1, \tag{4.16}$$

$$2)\ \text{If } (\sigma_2 - \sigma_3) \geq (\sigma_1 - \sigma_2),\ (\sigma_1 - \sigma_3) + (\sigma_2 - \sigma_3) = C$$

$$\text{or } \sigma_1 + \sigma_2 - 2\sigma_3 = C_2. \tag{4.17}$$

The constant, C_1, can be found by considering an x-direction tension test. At yielding, $\sigma_x = \sigma_1 = Y$, $\sigma_y = \sigma_z = \sigma_2 = \sigma_3 = 0$. Therefore, $(\sigma_1 - \sigma_2) > (\sigma_2 - \sigma_3)$, so criterion (1) applies, and $C = (\sigma_1 - \sigma_3) + (\sigma_1 - \sigma_2) = 2Y$. Therefore, $C_1 = 2Y$.

The constant, C_2, can be found by considering an x-direction compression test. At yielding, $\sigma_x = \sigma_3 = -Y$, $\sigma_y = \sigma_z = \sigma_2 = \sigma_3 = 0$. Therefore, $(\sigma_2 - \sigma_3) > (\sigma_1 - \sigma_2)$, so criterion 2 applies, and $C = (\sigma_1 - \sigma_3) + (\sigma_2 - \sigma_3) = -(-2Y)$ or $C_2 = 2Y$.

To make a plot, consider six loading paths (Figure 4.7).

In region A, $\sigma_x = \sigma_1$, $\sigma_y = \sigma_2$, $\sigma_z = \sigma_3 = 0$ and $\sigma_x > 2\sigma_y$ so $(\sigma_1 - \sigma_3) > (\sigma_1 - \sigma_2)$

Therefore, criterion (1), $(\sigma_x = 0) + (\sigma_x - \sigma_y) = 2Y$, or $\sigma_x = Y + \sigma_y/2$

In region B, $\sigma_x = \sigma_1$, $\sigma_y = \sigma_2$, $\sigma_z = \sigma_3 = 0$ but σ_y so $(\sigma_1 - \sigma_3) < (\sigma_1 - \sigma_2)$

Therefore, criterion (2), $(\sigma - 0) + (\sigma_y - 0) = 2Y$, or $\sigma_x = 2Y - \sigma_y$

In region C, $\sigma_y = \sigma_1$, $\sigma_x = \sigma_2$, $\sigma_z = \sigma_3 = 0$ but $\sigma_y < 2\sigma_x$ so $(\sigma_1 - \sigma_3) < (\sigma_1 - \sigma_2)$

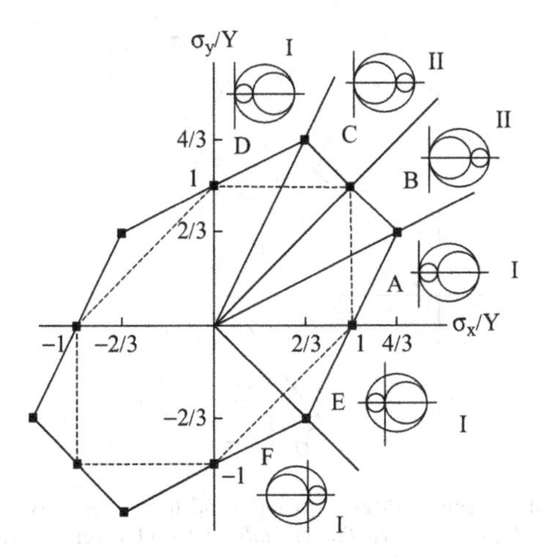

Figure 4.7. Another isotropic yield locus.

Therefore, criterion (2), $(\sigma_y - 0) + (\sigma_x = 0) = 2Y$, or $\sigma_y = 2Y - \sigma_x$

In region D, $\sigma_y = \sigma_1, \sigma_x = \sigma_2, \sigma_z = \sigma_3 = 0$ and $\sigma_y > 2\sigma_x$ so $(\sigma_1 - \sigma_2) > (\sigma_2 - \sigma_3)$

Therefore, criterion (1), $(\sigma_{y-0}\ 0) + (\sigma_y - \sigma_x) = 2Y$, or $\sigma_y = Y + \sigma_x/2$

In region E, $\sigma_x = \sigma_1, \sigma_y = \sigma_3, \sigma_z = \sigma_2 = 0$ and $(\sigma_1 - \sigma_2) > (\sigma_2 - \sigma_3)$

Therefore, criterion (1), $(\sigma_x - 0) + (\sigma_x - \sigma_y) = 2Y$, or $\sigma_x = Y + \sigma_y/2$

In region F, $\sigma_x = \sigma_1, \sigma_y = \sigma_3, \sigma_z = \sigma_2 = 0$ so $(\sigma_1 - \sigma_2) > (\sigma_2 - \sigma_3)$

Therefore, criterion (1), $(\sigma_x = 0) + (\sigma_x - \sigma_y) = 2Y$, or $\sigma_x = .Y + \sigma_y/2$

Plotting these in the appropriate regions (Figure 4.7), and using symmetry to construct the left hand half, this criterion extends the yield locus the maximum amount into the first quadrant, with $\sigma_x/Y = 4/3$.

EQUIVALENT STRESS STATES FOR ISOTROPY

For an isotropic material, the yield locus can be divided into three equivalent sectors as shown in Figure 4.8.

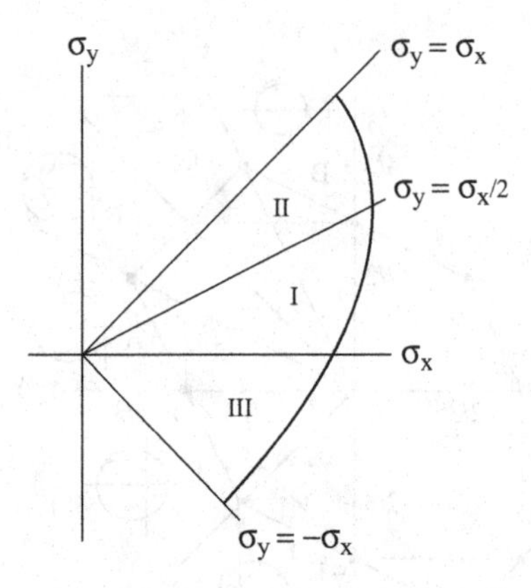

Figure 4.8. Equivalent regions of an isotropic yield locus. From W. F. Hosford, *The Mechanics of Crystals and Textured Polycrystals*, Oxford University Press (1993).

For a $\sigma_z = 0$ yield locus plot, the diameters of the Mohr's circles are $(\sigma_x - \sigma_z) > (\sigma_x - \sigma_y) > (\sigma_y - \sigma_z)$ in region I, $(\sigma_x - \sigma_z) > (\sigma_y - \sigma_z) > (\sigma_x - \sigma_y)$ in region II, and $(\sigma_x - \sigma_y) > (\sigma_x - \sigma_z) > (\sigma_z - \sigma_y)$ in region III.

PLOTTING ON THE π-PLANE

The π-plane is the plane in σ_1, σ_2, σ_3 space for which the reduced hydrostatic stress is zero. Yield loci plotted on the $\sigma_z = 0$ plane can be converted to plots on the π-plane. If the stresses on the $\sigma_z = 0$ plane are σ_x and σ_y, the reduced stresses are $\sigma_i' = \sigma_i - \sigma_H$, where σ_H is the hydrostatic stress $\sigma_H = (\sigma_x + \sigma_y + \sigma_z)/3 = (\sigma_x + \sigma_y)/3$, so

$$\sigma_x' = (2/3)\sigma_x - (1/3)\sigma_y$$

$$\sigma_y' = (2/3)\sigma_y - (1/3)\sigma_x$$

$$\text{and } \sigma_z' = -(1/3)\sigma_x - (1/3)\sigma_y - (1/3)\sigma_z \qquad 4.18$$

Substituting $X = \sigma_x'$ and $Y = (1/\sqrt{3})\sigma_x' + (2/\sqrt{3})\sigma_y'$, equation 4.18 becomes

$$X = (2/3)\sigma_x' - (1/3)\sigma_y' \quad \text{and} \quad Y = (1/\sqrt{3})\sigma_y'. \qquad 4.19$$

These relations are shown in Figure 4.9.

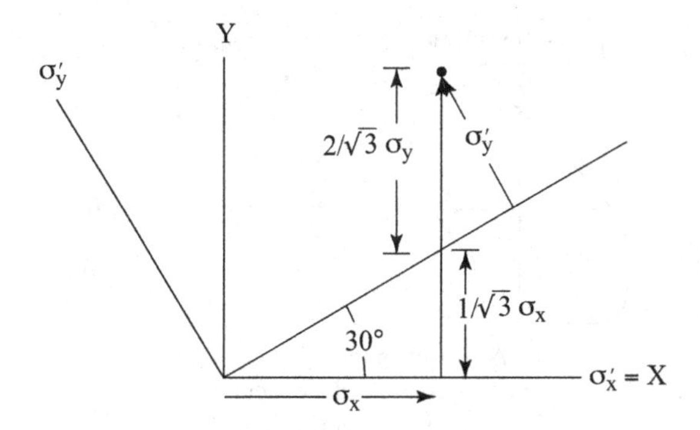

Figure 4.9. Coordinates for π-plane plotting. From W. F. Hosford, *The Mechanics of Crystals and Textured Polycrystals*, Oxford University Press (1993).

On the π-plane, the von Mises criterion plots as a circle and Tresca as a regular hexagon (Figure 4.10).

BASIC ASSUMPTIONS

Figure 4.11 shows yield loci that violate the basic assumptions of (1) isotropy, (2) independence of pressure, and (3) independence on the sign of external stresses.

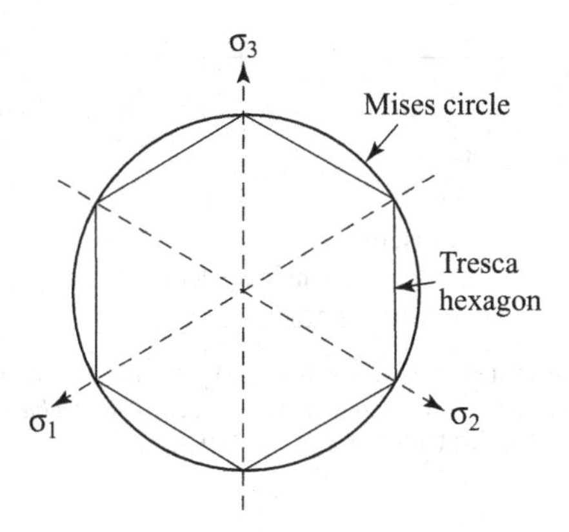

Figure 4.10. Plot of the Tresca and von Mises yield criteria on the pi plane. From W. F. Hosford and R. M. Caddell, *Metal Forming: Mechanics and Metallurgy*, 4th ed., Cambridge University Press (2011).

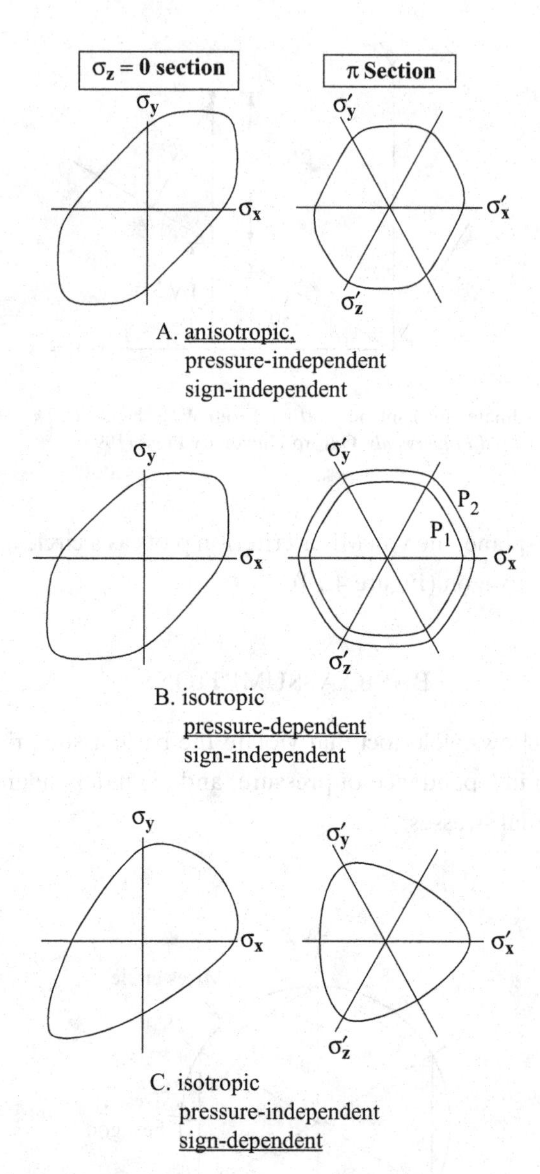

Figure 4.11. Yield loci that are (A) anisotropic, (B) pressure dependent, and (C) dependent on the sign of the stress state. From W. F. Hosford, *The Mechanics of Crystals and Textured Polycrystals*, Oxford University Press (1993).

Flow Rules

With elastic deformation, the strains are given by Hooke's law. There are similar relations for plastic deformation. When a material yields, the ratio of the resulting strains depends on the stress state that causes yielding. The general relations between plastic strains and the stress states are called the *flow rules*. They may be expressed as

$$d\varepsilon_{ij} = d\lambda(\partial f/\partial\sigma_{ij}), \qquad 4.20$$

where f is the yield function, which corresponds to the yield criterion of concern and $d\lambda$ is a constant that depends on the shape of the stress strain curve. For the von Mises criterion, if we write $f = [(\sigma_2 - \sigma_3)^2 + (\sigma_3 - \sigma_1)^2 + (\sigma_1 - \sigma_2)^2]/4$, then equation 4.20 results in

$$\begin{aligned} d\varepsilon_1 &= d\lambda[2(\sigma_1 - \sigma_2) - 2(\sigma_3 - \sigma_1)]/4 = dl[\sigma_1 - (\sigma_2 + \sigma_3)/2] \\ d\varepsilon_2 &= d\lambda[\sigma_2 - (\sigma_3 + \sigma_1)/2] \\ d\varepsilon_3 &= d\lambda[\sigma_3 - (\sigma_1 + \sigma_2)/2]. \end{aligned} \qquad 4.21$$

These are known as the Levy-Mises [7] equations. Even though $d\lambda$ is not usually known, these equations are useful for finding the ratio of strains that result from a known stress state or the ratio of stresses of stresses that correspond to a known strain state. The constant, $d\lambda$, can be expressed as $d\lambda = d\bar{\varepsilon}/d\bar{\sigma}$, which is the inverse slope of the effective stress-strain curve at the point where the strains are being evaluated.

It is interesting to note that equation 4.21 parallels Hooke's laws, where $d\lambda = d\bar{\varepsilon}/d\bar{\sigma}$ replaces $1/E$ and $1/2$ replaces Poisson's ratio, υ. For this reason, it is sometimes said that the "plastic Poisson's ratio" is 1/2.

For the Tresca yield criterion, the flow rules can be found by applying equation 4.20 with $f = \sigma_1 - \sigma_3$. Then $d\varepsilon_1 = d\lambda$, $de_2 = 0$, and $d\varepsilon = -d\lambda$, so

$$d\varepsilon_1 : d\varepsilon_2 : d\varepsilon_3 = 1 : 0 : -1 \qquad 4.22$$

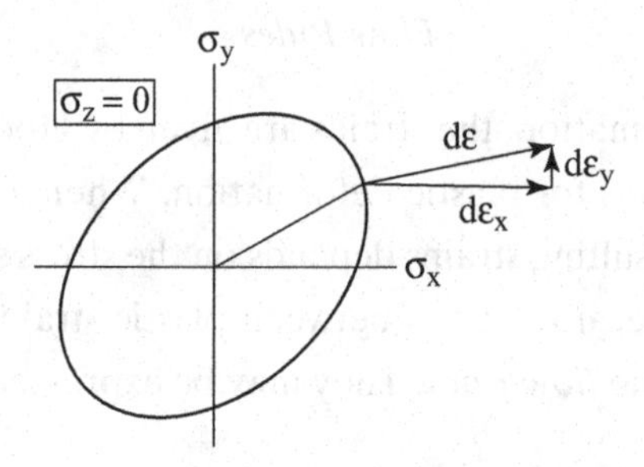

Figure 4.12. The ratios of strain resulting from yielding are in the same proportion as the components of the normal to the yield locus at the point of yielding. From W. F. Hosford, *The Mechanics of Crystals and Textured Polycrystals*, Oxford University Press (1993).

With the high exponent criterion, (4.14), the strains are in the ratio

$$d\varepsilon_1 : d\varepsilon_2 : d\varepsilon_3 = (\sigma_1-\sigma_3)^{a-1} + (\sigma_1-\sigma_2)^{a-1} : (\sigma_2-\sigma_3)^{a-1} + (\sigma_2-\sigma_1)^{a-1} : (\sigma_3-\sigma_1)^{a-1} + (\sigma_3-\sigma_2)^{a-1} \quad 4.23$$

PRINCIPLE OF NORMALITY

The *principle of normality* is a corollary to the flow rules. According to this principle, the strains that result from yielding are in the same ratio as the ratio of stress components of normal to the yield locus. This is illustrated in Figure 4.12. Figure 4.13 illustrates normality for several

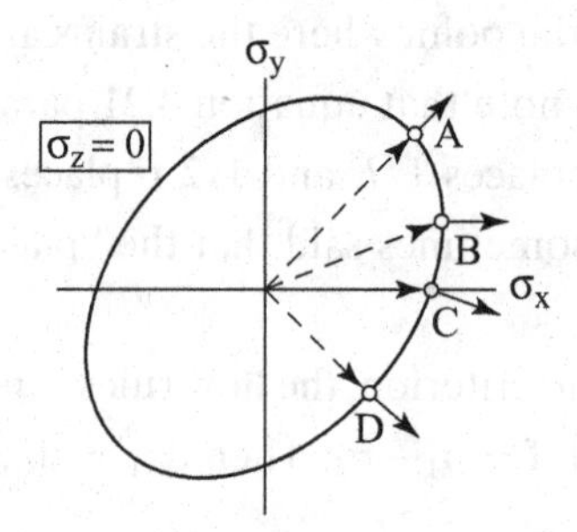

Figure 4.13. The ratio of strains along several simple loading paths. At A, $\varepsilon_y = \varepsilon_x$. At B, $\varepsilon_y = 0$. At C, $\varepsilon_y = -1/2\varepsilon_x$ At D, $\varepsilon_y = -\varepsilon_x$ From W. F. Hosford, *The Mechanics of Crystals and Textured Polycrystals*, Oxford University Press (1993).

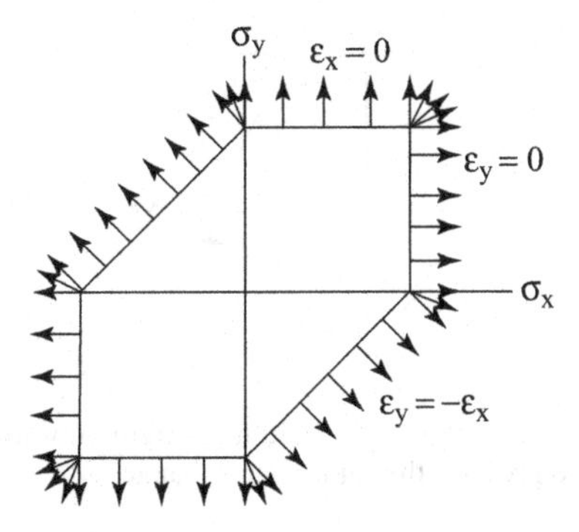

Figure 4.14. The normality principle applied to the Tresca yield criterion. All stress states on the same side of the locus cause the same shape change. The shape changes at the corners are ambiguous. Biaxial tension can produce any shape change from $\varepsilon_y = 0$ to $\varepsilon_x = 0$. From W. F. Hosford, *The Mechanics of Crystals and Textured Polycrystals*, Oxford University Press (1993).

simple loading paths. A corollary to equation 4.18 is that for a σ_1 vs. σ_2 yield locus with $\sigma_3 = 0$,

$$d\varepsilon_1/d\varepsilon_2 = -\partial\sigma_2/\partial\sigma_1, \qquad 4.24$$

where $\partial\sigma_2/\partial\sigma_1$ is the slope of the yield locus at the point of yielding. It should be noted that equation 4.20 is general and can be used with other yield criteria even those which account for anisotropy and for pressure-dependent yielding.

Figure 4.14 illustrates the normality principle normality applied to the Tresca yield criterion. All stress states except those corresponding to the corners of the yield locus result in plane strain. The shape change resulting from stress states at the corners are ambiguous.

CONVEXITY

A corollary to the normality principle is that a yield locus cannot be outwardly concave. This is illustrated by Figure 4.15, which shows that if a yield locus were outwardly concave, two different stress states would produce the same shape change.

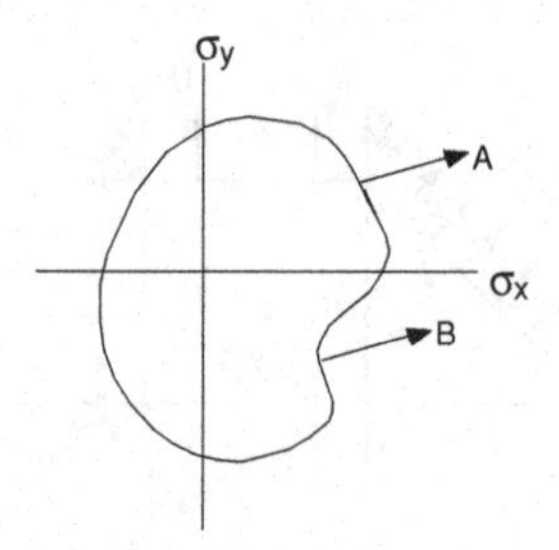

Figure 4.15. Illustration showing that if a yield locus were outwardly concave, two different stress states would produce the same shape change.

Figure 4.15 shows that if a yield locus were outwardly concave, two different stress states would produce the same shape change.

LODE VARIABLES

Lode [4] suggested that two variables, $\nu = 2(2\rho + 1)/(2\rho) - 1$, which describes the state of strain, and $\mu = 3\alpha - 1$, which describes the state of stress, could be used to investigate yield criteria. Taylor and Quinney [5] made tests on tube of a number of materials loaded simultaneously under tension and torsion. Figure 4.16 shows their results, which clearly lie between the predictions of the von Mises and the Tresca criteria. The predictions of the high exponent criterion with a = 6 are a better approximation of the data.

Effective Stress and Effective Strain

The concepts of effective stress and effective strain are useful in analyzing the strain hardening that occurs on loading paths other than uniaxial tension. Effective stress, $\bar{\sigma}$, and effective strain, $\bar{\varepsilon}$, are defined so that:

1) $\bar{\sigma}$ and $\bar{\varepsilon}$ reduce to σ_x and ε_x in an x-direction tension test.
2) The incremental work per volume done in deforming a material plastically is $dw = \bar{\sigma} d\bar{\varepsilon}$.

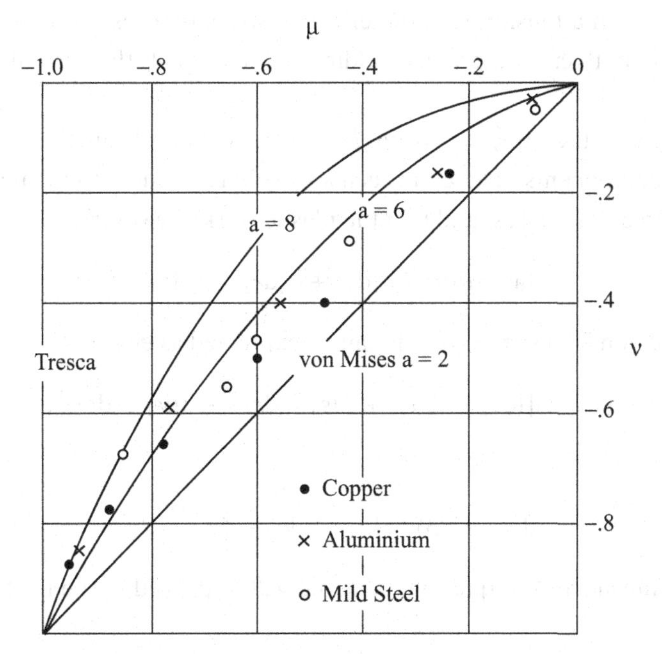

Figure 4.16. Lode variables found experimentally by Taylor and Quinney, together with the predictions of the high exponent criterion, equations 4.14 and 4.21.

3) Furthermore, it is usually assumed that the $\bar{\sigma}$ vs. $\bar{\varepsilon}$ curve describes the strain hardening for loading under a constant stress ratio, α, regardless of α, so $\bar{\sigma} = \mathrm{f}(\bar{\varepsilon})$.

With large strains, the $\bar{\sigma}$ vs. $\bar{\varepsilon}$ curves do depend on the loading path because of orientation changes within the material, which is dependent on the loading path. However, the dependence of strain hardening on loading path is significant only at large strains.

The effective stress, $\bar{\sigma}$, is the function of the applied stresses that determine whether yielding occurs. When $\bar{\sigma}$ reaches the current flow stress, plastic deformation will occur. For the von Mises criterion,

$$\bar{\sigma} = (1/\sqrt{2})\left[(\sigma_2 - \sigma_3)^2 + (\sigma_3 - \sigma_1)^2 + (\sigma_1 - \sigma_2)^2\right]^{1/2}, \qquad 4.25$$

and for the Tresca criterion,

$$\bar{\sigma} = (\sigma_1 - \sigma_3). \qquad 4.26$$

Note that in a tension test, the effective stress reduces to the tensile stress so both criteria predict yielding when $\bar{\sigma}$ equals the current flow stress.

Because the effective strain, $\bar{\varepsilon}$, is a mathematical function of the strain components, defined in such a way that $\bar{\varepsilon}$ reduces to the tensile strain in a tension test and that the plastic work per volume,

$$dw = \bar{\sigma}d\bar{\varepsilon} = \sigma_1 d\varepsilon_1 + \sigma_2 d\varepsilon_2 + \sigma_3 d\varepsilon_3. \quad 4.27$$

For the von Mises criterion, $d\bar{\varepsilon}$ can be expressed as either

$$d\bar{\varepsilon} = (\sqrt{2/3})[(d\varepsilon_2 - d\varepsilon_3)^2 + (d\varepsilon_3 - d\varepsilon_1)^2 + (d\varepsilon_1 - d\varepsilon_2)^2]^{1/2}, \quad 4.28$$

or as

$$d\bar{\varepsilon} = (2/3)^{1/3}\left(d\varepsilon_1^2 + d\varepsilon_2^2 + d\varepsilon_3^2\right)^{1/2}. \quad 4.29$$

The equivalence of equations 4.28 and 4.29 is proved in the next section.

Realizing that in a 1-direction tension test $d\varepsilon_2 = d\varepsilon_3 = -(1/2)d\varepsilon_1$, it can be shown that $d\bar{\varepsilon}$ in equations 4.28 and 4.29 reduce to $d\varepsilon_1$ in a 1-direction tension test by substituting.

$$d\bar{\varepsilon} = (\sqrt{2}/3)\{(0)^2 + [-(1/2)d\varepsilon_1 - d\varepsilon_1]^2 + [d\varepsilon_1 - (-1/2)d\varepsilon_1]^2\}1/2$$
$$= (\sqrt{2}/3)\left[(9/4)d\varepsilon_1^2 + (9/4)d\varepsilon_1^2\right]^{1/2} = d\varepsilon_1 \quad 4.30$$

If the straining is proportional (constant ratios of $d\varepsilon 1$: $d\varepsilon 2$: $d\varepsilon 3$), the total effective strain can be expressed as

$$\bar{\varepsilon} = [(2/3)(\varepsilon_1^2 + \varepsilon_2^2 + \varepsilon_3^2)]^{1/2}. \quad 4.31$$

If the straining is not proportional, $\bar{\varepsilon}$ must be found by integrating $d\bar{\varepsilon}$ along the strain path.

The effective strain (and stress) may also be expressed in terms of non-principal strains (and stresses). For von Mises,

$$\bar{\varepsilon} = \left[(2/3)\left(\varepsilon_x^2 + \varepsilon_y^2 + \varepsilon_z^2\right) + (1/3)\left(\gamma_{yz}^2 + \gamma_{zx}^2 + \gamma_{xy}^2\right)\right]^{1/2}, \quad 4.32$$

and

$$\bar{\sigma} = (1/\sqrt{2})\left[(\sigma_y - \sigma_z)^2 + (\sigma_z - \sigma_x)^2 + (\sigma_x - \sigma_y)^2 + 6\left(\tau_{yz}^2 + \tau_{zx}^2 + \tau_{xy}^2\right)\right]^{1/2}. \quad 4.33$$

DERIVATION OF THE VON MISES EFFECTIVE STRAIN [8]

The effective strain, $\bar{\varepsilon}$, is defined so that the incremental plastic work per volume.

$$dw = \sigma_1 d\varepsilon_1 + \sigma_2 d\varepsilon_2 + \sigma_3 d\varepsilon_3 = \bar{\sigma} d\bar{\varepsilon} \quad 4.34$$

where $\bar{\sigma}$ is the effective stress. For simplicity, consider a state of plane stress, $\sigma_3 = 0$. Then,

$$\bar{\sigma} d\bar{\varepsilon} = \sigma_1 d\varepsilon_1 + \sigma_2 d\varepsilon_2 = \sigma_1 d\varepsilon_1 (1 + \alpha\rho) \quad 4.35$$

where $\alpha = \sigma_2/\sigma_1$ and $\rho = d\varepsilon_2/d\varepsilon_1$, so

$$d\bar{\varepsilon} = d\varepsilon_1 (1 + \alpha\rho) \quad 4.36$$

From the flow rules (equation 4.21),

$\rho = d\varepsilon_2/d\varepsilon_1 = [\sigma_2 + (1/2)\sigma_1]/[\sigma_1 + (1/2)\sigma_2] = (2\alpha - 1)/(1 - 2\alpha)$ or
$\alpha = (2\rho + 1)(2 + \rho)$ 4.37

Combining equations 4.35 and 4.36,

$$d\bar{\varepsilon} = d\varepsilon_1 (\sigma_1/\bar{\sigma})[2(1 + \rho + \rho^2)/(2 + \rho)]. \quad 4.38$$

With $\sigma_3 = 0$, the von Mises effective stress is

$$\bar{\sigma} = \left(\sigma_1^2 + \sigma_2^2 - \sigma_1\sigma_2\right)^{1/2} = (1 - \alpha - \alpha^2)^{1/2}. \quad 4.39$$

Combining equations 4.37 and 4.39,

$$\sigma_1/\bar{\sigma} = ([(2 + \rho)/\sqrt{3}]/\sqrt{(}1 + \rho + \rho^2). \quad 4.40$$

Since $\rho = d\varepsilon_2/d\varepsilon_1$,

$$d\bar{\varepsilon} = (2/\sqrt{3})\left(d\varepsilon_1^2 + d\varepsilon_1 d\varepsilon_2 + d\varepsilon_2^2\right)^{1/2}. \qquad 4.41$$

Realizing that $d\varepsilon_1^2 + d\varepsilon_1 d\varepsilon_2 + d\varepsilon_2^2 = d\varepsilon_1^2 + d\varepsilon_1 d\varepsilon_2 + (-d\varepsilon_1^2 - d\varepsilon_2^2) = 2(d\varepsilon_1^2 + d\varepsilon_1 d\varepsilon_2 + d\varepsilon_2^2)$, equation 4.41 becomes

$$d\bar{\varepsilon} = (2/3)^{1/2}\left(d\varepsilon_1^2 + d\varepsilon_2^2 + d\varepsilon_3^2\right)^{1/2}. \qquad 4.42$$

This derivation is valid even if $\sigma_3 \neq 0$, because the stress state, $\sigma_3 = 0$, is equivalent to

$$\sigma_1' = \bar{\sigma} = \sigma_3,\ \sigma_2' = \sigma_2 - \sigma_3,\ \sigma_3' = \sigma_1 - \sigma_3 = 0.$$

TRESCA EFFECTIVE STRAIN

For the Tresca criterion, the effective strain is the absolutely largest principal strain,

$$d\bar{\varepsilon}_{Tresca} = |d\varepsilon_i|_{max}. \qquad 4.43$$

Although the Tresca effective strain is not widely used, it is of value because it is so extremely simple to find. It is worth noting because the von Mises effective strain can never differ greatly from it. Always

$$|d\varepsilon_i|_{max} \leq d\bar{\varepsilon}_{Mises} \leq 1.15|d\varepsilon_i|_{max}. \qquad 4.44$$

GENERAL EFFECTIVE STRAINS

Thc effective stress for Hill's 1948 criterion is

$$\bar{\sigma} = \{[R(\sigma_y - \sigma_z)^2 + P(\sigma_z - \sigma_x)^2 + RP(\sigma_x - \sigma_y)^2/[P(R+1]\}^{1/2}, \qquad 4.45$$

which reduces to σ_y for uniaxial tension in the x-direction. The corresponding expression of effective strain is

$$\bar{\varepsilon} = C[P(\varepsilon_y - R\varepsilon_z)^2 + R(P\varepsilon_z - \varepsilon_x)^2 + (R\varepsilon_x - P\varepsilon_y)^2]^{1/2} \qquad 4.46$$

with $C = [(R+1)/R]^{1/2}/(R+P+1)$

The effective strain expression for any yield criterion can be derived by setting
$\bar{\sigma}\bar{\varepsilon} = \sigma_{x'}\varepsilon_{x'} + \sigma_{y'}\varepsilon_{y'} + \sigma_{z'}\varepsilon_{z'}$ with $\sigma_{z'} = 0$ and substituting $\sigma_{y'} = \alpha\sigma_{x'}$ and $\varepsilon_{y'} = -\rho\varepsilon_{x'}$. This can be written as $\bar{\varepsilon}/\varepsilon_{x'} = [1/(\bar{\sigma}/\sigma_{x'})]/(1-\alpha\rho)$. It should be noted that with nonquardatic yield criteria, α cannot be solved directly from ρ. It must be found by trial and error.

As a final note, it must be emphasized that the effective strain should correpond to the yield criterion.

NOTE OF INTEREST

Christian Otto Mohr was born October 8, 1835 in the Holstein region of Germany and died October 2, 1918. At the age of 16, he attended the Polytechnic University in Hanover. Much of his early career was spent building railroads and designing bridges. He utilized some of the earliest steel trusses. As a result of his work, he became interested in mechanics. In 1867, he became a professor of mechanics at Stuttgart Polytechnic and in 1873 at Dresden Polytechnic. His direct and unpretentious lecturing style was popular with students. He developed the method for visually representing three dimensional stress states. In 1882, he developed the graphical method known as Mohr' circles or analyzing stress, and used it to propose an early theory of strength based on shear stress. He retired in 1900, yet continued his scientific work in the town of Dresden until his death on October 2, 1918.

Henri Édouard Tresca was born in France on October 12, 1814. He was a mechanical engineer and a professor at the Conservatoire National des Arts et Métiers in Paris. He is often regarded as the father of the field of plasticity, which he explored in an extensive series of experiments that began in 1864, when he postulated the Tresca yield criterion. Tresca's stature as an engineer was such that Gustave Eiffel put his name on number three in a list of people who made the Eiffel Tower in Paris possible. Tresca was also among the designers of the standard meter etalon. These bars had a modified cross section named for Tresca who designed them. The Tresca section was designed to provide maximum stiffness. Tresca was made an honorary member of the American Society of Mechanical Engineers in 1882. He died on June 21, 1885.

Richard Edler von Mises was born on April 19, 1883 in Lviv Poland. He worked on solid mechanics, fluid mechanics, aerodynamics statistics, and probability theory. He held the position of Gordon-McKay Professor of Aerodynamics and Applied Mathematics at Harvard University. In 1913, he postulated the von Mises yield criterion. He died on July 14, 1953 in Boston.

REFERENCES

1. H. Tresca, *Comptes Rendus Acad. Paris* v. 59 (1864).
2. R. Von Mises, *Göttin. Nachr. Math. Phys.* v. 1 (1913).
3. W. Lode, *Z. Phys* v. 36 (1926).
4. G. I. Taylor and H. Quinney, *Phil. Trans. Roy. Soc A.* v. 230 (1931).
5. B. Paul, in *Fracture, An Advanced Treatise, Mathemetical Fundamentals*, Liebowitz ed, v. Academic Press (1968).
6. W. F. Hosford, *J. Appl. Mech. (Trans. ASME ser E.)* v. 39E (1972).
7. M. Levy, *Comptes Rendus Acad. Paris* v. 7 (1870).
8. W. F. Hosford and R. M. Caddell, *Metal Forming: Mechanics and Metallurgy*, 4th ed, Cambridge University Press (2011).

5

BOUNDING THEOREMS AND WORK PRINCIPLES

Calculation of exact forces to cause plastic deformation in metal forming processes is often difficult. Exact solutions must be both *statically* and *kinematically* admissible. This means they must be geometrically self-consistent as well as satisfying stress equilibrium everywhere in the deforming body. Slip-line field analysis for plane strain deformation satisfies both and are therefore exact solutions. This topic is treated in Chapter 15. Upper and lower bounds are based on well-established principles [1, 2].

Frequently, it is difficult to make exact solutions and it is simpler to use limit theorems, which allows one to make analyses that result in calculated forces that are known to be either correct or too high or too low than the exact solution.

UPPERBOUNDS

The upper bound theorem states that any estimate of the forces to deform a body made by equating the rate of internal energy dissipation to the external forces will equal or be greater than the correct force. The analysis involves:

1. Assuming an internal flow field that will produce the shape change.
2. Calculating the rate at which energy is consumed by this flow field.
3. Calculating the external force by equating the rate of external work with the rate of internal energy consumption.

The flow field can be checked for consistency with a velocity vector diagram or *hodograph*. In the analysis, the following simplifying assumptions are usually made:

1. The material is homogeneous and isotropic.
2. There is no strain hardening.
3. Interfaces are either frictionless or sticking friction prevails.
4. Usually only two-dimensional (plane-strain) cases are considered with deformation occurring by shear on a few discrete planes. Everywhere else the material is rigid.

LOWER BOUNDS

Lower bounds are based on satisfying stress equilibrium, while ignoring geometric self-consistency. They give forces that are known to be either too low or correct. As such, they can assure that a structure is "safe." A lower bound solution to any problem requires a statically admissible stress field. There must be a balance of stresses across all boundaries. In metal forming calculations, stresses must balance across interfaces between work piece and tooling. In analysis of the deformation of polycrystalline materials, the stress states must be the same on both sides of a grain boundary.

PRINCIPLE OF MAXIMUM VIRTUAL WORK

The principle of maximum virtual work states that of all of the stress states that might cause yielding, the appropriate one for a given shape change is the one for which the calculated work is largest [1]. Two examples follow:

For a material that obeys the Tresca yield criterion with $\sigma_z = 0$ (Figure 4.2), the stress states that can cause yielding are:

1. $\sigma_x = +Y, 0 \le \sigma_y \le Y$
2. $\sigma_y = +Y, 0 \le \sigma_x \le Y$

3. $\sigma_x - \sigma_y = +Y$.
4. $\sigma_x = -Y, 0 \geq \sigma_y \geq Y$
5. $\sigma_y = +Y, 0 \geq \sigma_x \geq Y$
6. $\sigma_x - \sigma_y = -Y$.

The plastic work to cause axially symmetric flow about x, $d\varepsilon_y = d\varepsilon_y = -(1/2)d\varepsilon_x$ is $dw = \sigma_x d\varepsilon_x + \sigma_y d\varepsilon_x = (\sigma_x + \sigma_y/2)d\varepsilon_x$. For stress states 1 and 3, with $\sigma_y = 0$, the calculated work is $dw = Y$. For all of the other states, the calculated work is less, so states 1 and 3 are appropriate. For plane strain, $d\varepsilon_y = 0$, the calculated work for stress state 1 is $dw = Y$. For all of the other states, the calculated work is less, so state 1 is appropriate.

A frictional system provides a second example. Sliding occurs when the shearing force, F, in any direction in the plane of sliding reaches a value μN, where μ is the coefficient of friction and N is the normal force. For sliding in some direction, d, only the shear stress acting in the direction, d, is appropriate. The calculated work, $W = F \cdot d\cos\theta$, where θ is the angle between the direction of sliding and the direction of the force, is not a maximum if $\theta \neq 0$.

The principle of maximum virtual work can be regarded as a corollary to the normality principle. When a material yields, the resulting strains are in the ratio that maximizes the work done.

MINIMUM WORK THEOREM

Of all the stress states that could cause yielding, the one that requires the least total work is most appropriate. The concept of a minimum work principle seems to contradict the principle of maximum virtual work, but it does not. Consider again a frictional system. Sliding in many directions could achieve a displacement x_d in the d direction. This displacement could occur in one step by a force F_d or it could occur by two displacements, $(1/2)x_d/\cos\theta$ under a force F and a second displacement, $(1/2)x_d(-\cos\theta)$ under a force F. Clearly, the work done by the single displacement is lower and appropriate.

NOTE OF INTEREST

Daniel Charles Drucker was born June 3, 1918 and died of leukemia September 1, 2001. He was an international authority on the theory of plasticity. He taught at Brown University from 1946 until 1968 when he joined the University of Illinois as Dean of Engineering. He was awarded the Timoshenko Medal in 1983. In 1984, he left Illinois to become a graduate research professor at the University of Florida until his retirement in 1994. In 1988, he was awarded the National Medal of Science. He was a member of the National Academy of Engineering and of the American Academy of Arts and Sciences.

REFERENCES

1. D. C. Drucker, H. J. Greenberg, and W. Prager. *J. Appl. Mech.* v. 18 (1951).
2. R. Hill, *Phil Mag*. v. 42 (1951).

6

SLIP-LINE FIELD THEORY

INTRODUCTION

Slip-line field analysis involves plane-strain deformation fields that are both geometrically self-consistent and statically admissible. Therefore, the results are exact solutions. *Slip lines*[1] are really planes of maximum shear stress and are oriented at 45 degrees to the axes of principal stress. The basic assumptions are that the material is isotropic and homogeneous and rigid-ideally plastic (that is, no strain hardening and that shear stresses at interfaces are constant). Effects of temperature and strain rate are ignored.

Figure 6.1 shows a very simple slip-line field for indentation. In this case, the thickness, t, equals the width of the indenter, b and both are very much smaller than w. The maximum shear stress occurs on lines DEB and CEA. The material in triangles DEA and CEB is rigid. Although the field must change as the indenters move closer together, the force can be calculated for the geometry as shown. The stress, σ_y, must be zero because there is no restrain to lateral movement. The stress, σ_z, must be intermediate between σ_x and σ_y. Figure 6.2 shows the Mohr's circle for this condition. The compressive stress necessary for this indentation, $\sigma_x = -2k$. Few slip-line fields are composed of

[1] The term *slip lines* as used here should not be confused with the microscopic slip lines found on the surface of crystals undergoing crystallographic slip.

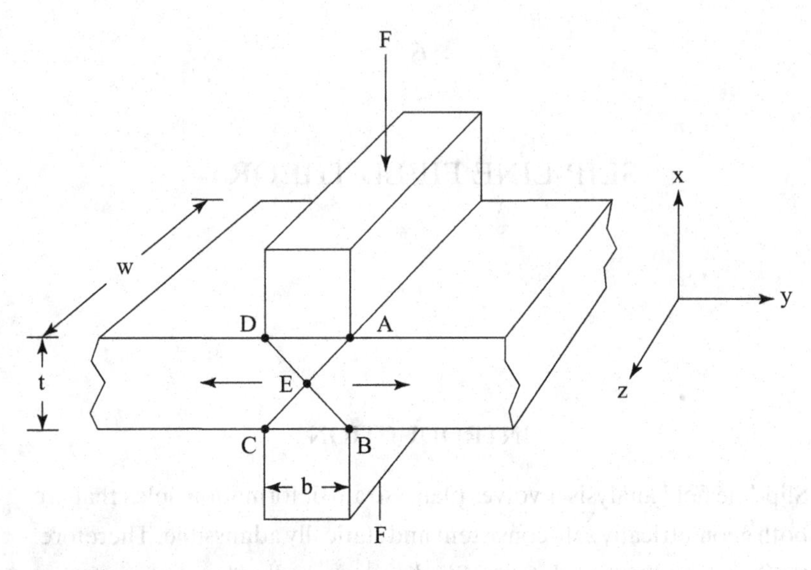

Figure 6.1. A slip-line field for frictionless plane-strain indentation. From W. F. Hosford and R. M. Caddell, *Metal Forming: Mechanics and Metallurgy*, 4th ed. Cambridge University Press (2011).

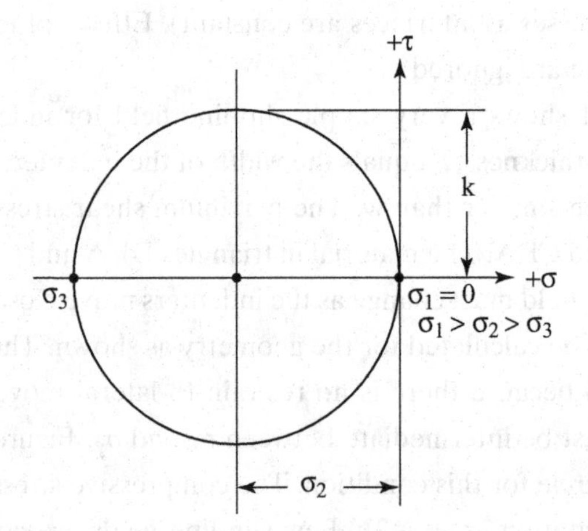

Figure 6.2. Mohr's stress circle for frictionless plane-strain indentation in Figure 6.1. From W. F. Hosford and R. M. Caddell, *Metal Forming: Mechanics and Metallurgy*, 4th ed. Cambridge University Press (2011).

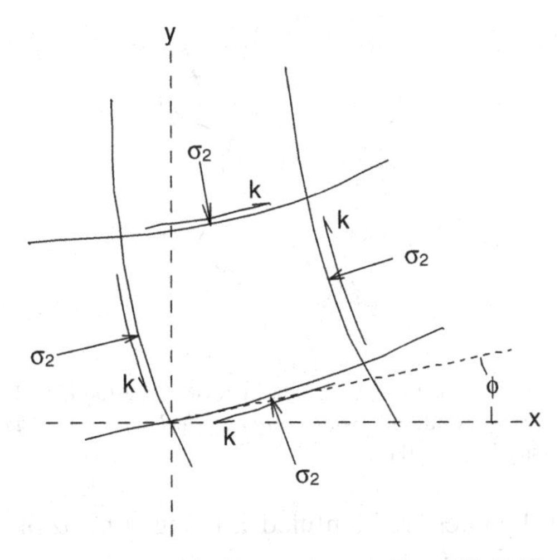

Figure 6.3. Stresses acting on a curvilinear element. From W. F. Hosford and R. M. Caddell, *Metal Forming: Mechanics and Metallurgy*, 4th ed. Cambridge University Press (2011).

only straight lines. More complicated fields are considered throughout this chapter.

GOVERNING STRESS EQUATIONS

With plane strain ($d\varepsilon_2 = 0$) all of the flow is in the x-y plane, so $d\varepsilon_3 = -d\varepsilon_1$ and $\sigma_2 = \left(\sigma_x + \sigma_y\right)/2$. Therefore, according to the von Mises criterion, σ_2 is always the mean or hydrostatic stress.

$$\sigma_2 = (\sigma_1 + \sigma_2 + \sigma_3)/3 = \sigma_{\text{mean}} \quad 6.1$$

$$\text{and } \sigma_1 = \sigma_2 + k,\ \sigma_3 = \sigma_2 - k. \quad 6.2$$

As such, plane-strain deformation can be considered as pure shear with a super-imposed hydrostatic stress, σ_2.

The planes of maximum shear stress are mutually perpendicular and form a series of orthogonal slip lines (Figure 6.3). The shear stress acting on these lines is k, whereas the mean stress, σ_2, acts perpendicular to the slip lines. The slip lines are rotated at some angle, ϕ, to the x and y axes.

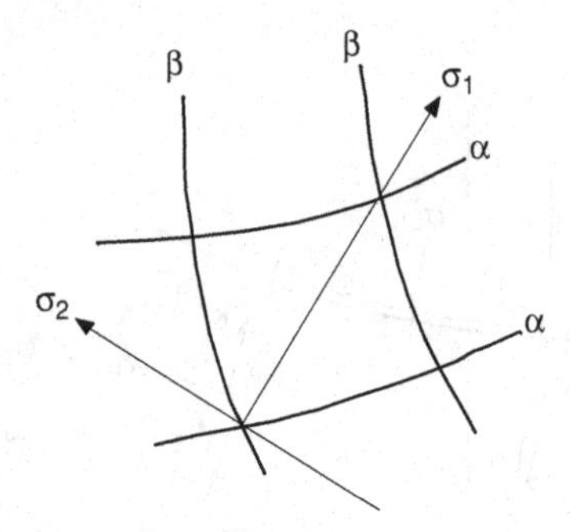

Figure 6.4. The 1-axis lies in the first quadrant formed by the α and β lines. From W. F. Hosford and R. M. Caddell, *Metal Forming: Mechanics and Metallurgy*, 4th ed. Cambridge University Press (2011).

Families of slip lines are identified as being either α or β lines. The convention is that the largest principal stress, σ_1, (most tensile) lies in the first quadrant formed by α and β lines as illustrated in Figure 6.4.

For plane strain, τ_{xy} and τ_{zx} are zero, so the equilibrium equations reduce to

$$\partial\sigma_x/\partial x + \partial\tau_{yx}/\partial y = 0 \quad \text{and } \partial\sigma_y/\partial y + \partial\tau_{xy}/\partial x = 0. \tag{6.3}$$

From the Mohr's stress circle diagram, Figure 6.5,

$$\sigma_x = \sigma_2 - 2k\sin\phi, \quad \sigma_y = \sigma_2 + 2k\sin\phi, \quad \tau_{xy} = k\cos\phi. \tag{6.4}$$

Differentiating equation 5.4 and substituting into equation 6.3,

$$\partial\sigma_2/\partial x - 2k\cos 2\phi\partial\phi/\partial x - 2k\sin 2\phi\partial\phi/\partial y = 0 \text{ and } \partial\sigma_2/\partial y + 2k\cos 2\phi\partial\phi/\partial y - 2k\sin 2\phi\partial\phi/\partial x = 0. \tag{6.5}$$

A set of axes, x′ and y′, can be oriented so that they are tangent to the α and β lines at the origin. In that case, $\phi = 0$. In this case, equation 6.5 reduces to

$$\partial\sigma_2/\partial x' - 2k\partial\phi/\partial x' = 0 \quad \text{and } \partial\sigma_2/\partial y' - 2k\partial\phi/\partial y' = 0. \tag{6.6}$$

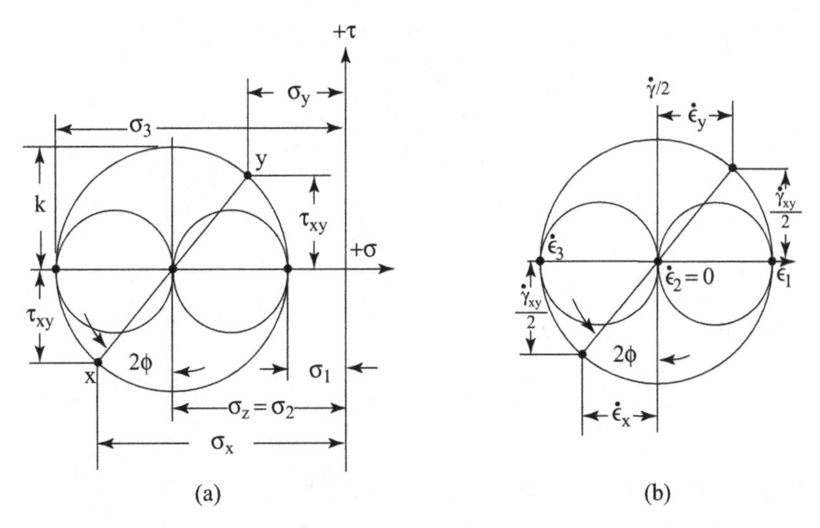

Figure 6.5. (a) Mohr's stress and (b) strain-rate circle for plane-strain. From W. F. Hosford and R. M. Caddell, *Metal Forming: Mechanics and Metallurgy*, 4th ed. Cambridge University Press (2011).

Integrating,

$$\sigma_2 = 2k\phi + C_1 \text{ along an } \alpha \text{ line and}$$
$$\sigma_2 = -2k\phi + C_2 \text{ along an } \beta \text{ line.} \qquad 6.7$$

Moving along an α line or a β line causes σ_2 to change by

$$\Delta\sigma_2 = 2k\Delta\phi \text{ along an } \alpha \text{ line}$$
$$\text{and } \Delta\sigma_2 = -2k\Delta\phi \text{ along a } \beta \text{ line.} \qquad 6.8$$

If σ_2 is replaced by $-P$ (pressure) equations 6.8 and 6.9 are written as

$$\Delta P = -2k\Delta\phi \text{ along an } \alpha \text{ line}$$
$$\Delta P = +2k\Delta\phi \text{ along an } \beta \text{ line} \qquad 6.9$$

BOUNDARY CONDITIONS

The direction of one principal stress can be found at a boundary. The force and stress normal to a free surface is a principle stress, so the α- and β-lines must meet the surface at 45 degrees. The α and β lines

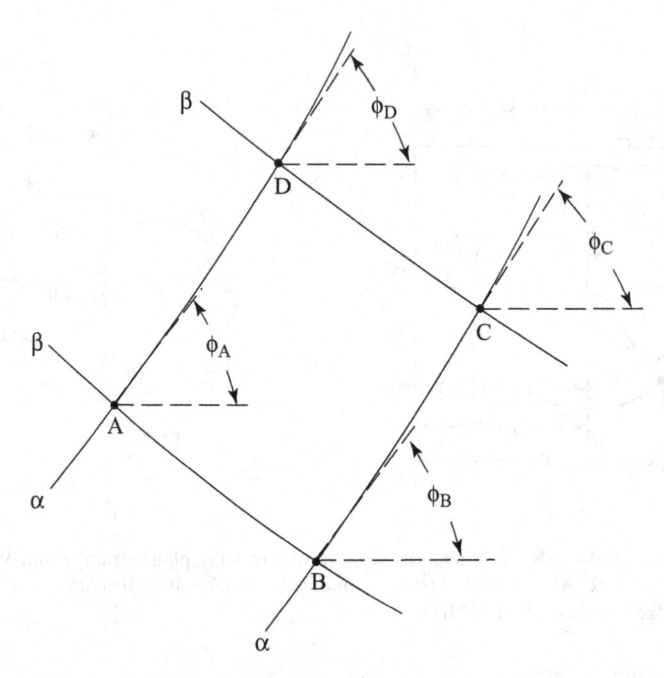

Figure 6.6. Two pairs of α- and β-lines for analyzing the change in mean normal stress by traversing two different paths. From W. F. Hosford and R. M. Caddell, *Metal Forming: Mechanics and Metallurgy*, 4th ed. Cambridge University Press (2011).

must also meet a frictionless surface at 45 degrees and surfaces of sticking friction at 0 and 90 degrees.

According to equation 6.7, the difference between σ_2 at A and C in Figure 6.6 can be found by traversing either of two paths, ABC or ADC. On the path through B, $\sigma_{2B} = \sigma_{2A} - 2k(\phi_B$ 4.43. $- \phi_A)$ and $\sigma_{2C} = \sigma_{2B} + 2k(\phi_C - \phi_B) = \sigma_{2A} - 2k(2\phi_B - \phi_A - \phi_C)$. On the other hand, traversing the path ADC, $\sigma_{2D} = \sigma_{2A} + 2k(\phi_D - \phi_A)$ and $\sigma_{2C} = \sigma_{2D} - 2k(\phi_C - \phi_D) = \sigma_{2A} - 2k(2\phi_d - \phi_A - \phi_C)$. Comparing these two paths,

$$\begin{aligned} \phi_A - \phi_B &= \phi_D - \phi_C \text{ and} \\ \phi_A - \phi_D &= \phi_B - \phi_C. \end{aligned} \qquad 6.10$$

Equation 6.10 implies that the net of α and β lines must be such that the change of ϕ is the same along a family of lines moving from one

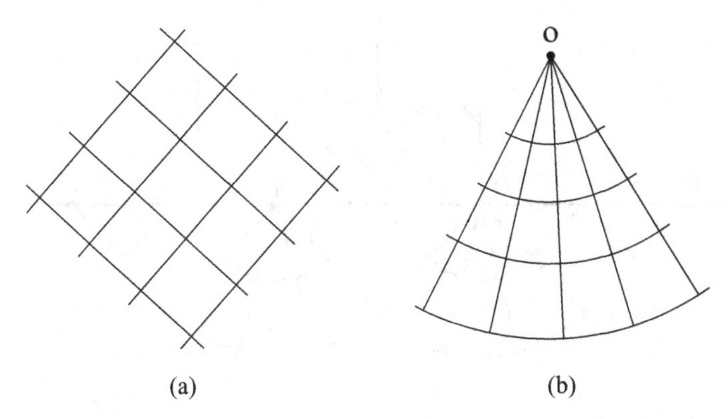

Figure 6.7. (a) Net of straight lines (b) Centered fan. From W. F. Hosford and R. M. Caddell, *Metal Forming: Mechanics and Metallurgy*, 4th ed. Cambridge University Press (2011).

intersection with the opposite family to the next intersection. This together with the orthogonality requirement indicates that any angular change along a line is of significance but the distance along the line is not.

There are two simple fields that meet these requirements. One is a set of straight lines (Figure 6.7a) and the other is a centered fan (Figure 6.7b). The mean stress, σ_2, is the same everywhere in the field of straight lines, so it is a constant pressure zone. In the centered-fan field, σ_2 is the same everywhere along a given radius but is different on different radii.

PLANE-STRAIN INDENTATION

A number of problems can be solved with these two fields. Consider plane-strain indentation. A possible field consisting of two centered fans and a constant pressure zone is shown in Figure 6.8. The α and β lines can be identified by realizing that the stress normal to OC is zero or alternatively, that under the indenter the most compressive stress is parallel to $P_\perp$.

A more detailed illustration of the field is given in Figure 6.9.

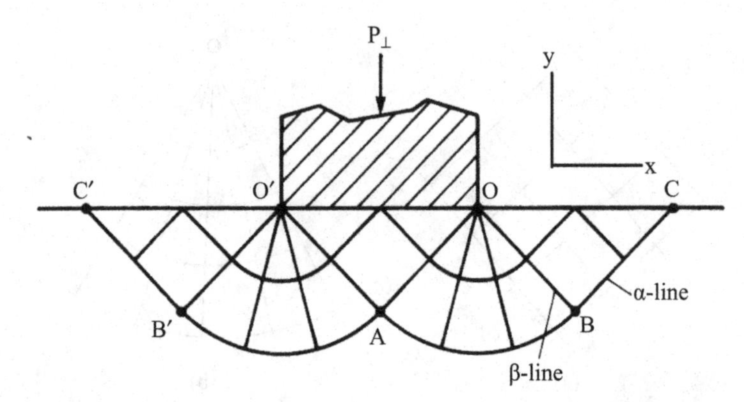

Figure 6.8. A possible slip-line field for plane-strain indentation. From W. F. Hosford and R. M. Caddell, *Metal Forming: Mechanics and Metallurgy*, 4th ed. Cambridge University Press (2011).

Along OC, $\sigma_y = \sigma_1 = 0$, $\sigma_2 = -\text{k}$, $\sigma_x = \sigma_3 = -2\text{k}$. Rotating clockwise along CBAO′ on the α line through $\Delta\phi = -\pi/2$, $\sigma'_{2\text{OO}} = \Delta_{2\text{OC}} + 2\text{k}\Delta\phi_\alpha = -\text{k} + 2\text{k}(-\pi/2)$,

$P_\perp/2k = P_\text{O}/2k = -\sigma_{2\text{OO}}/2k = 1 + \pi/2 = 2.57$. The pressure is constant but different in regions OBC and in O′OA. Although the

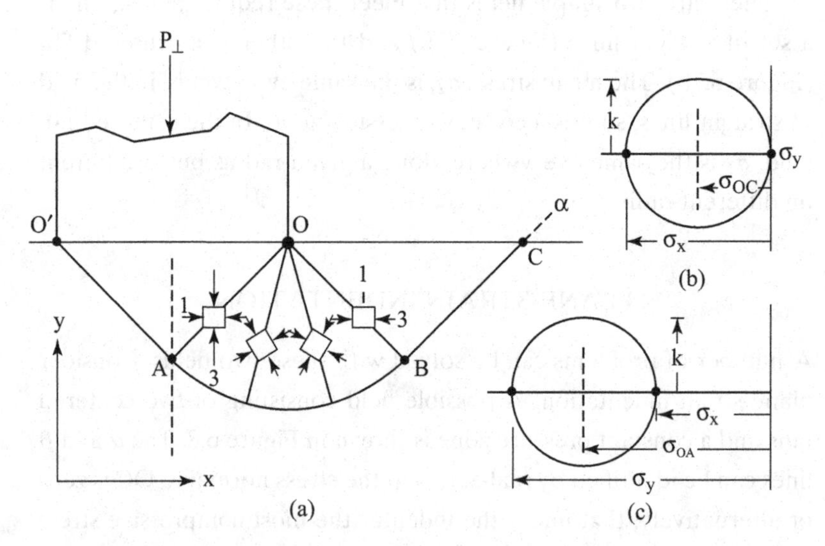

Figure 6.9. A detailed view of Figure 5.8 showing the changing stress state and the Mohr's stress circles for triangles OBC and O′OA. From W. F. Hosford and R. M. Caddell, *Metal Forming: Mechanics and Metallurgy*, 4th ed. Cambridge University Press (2011).

metal is stressed to its yield stress in these regions, it does not deform.

With the von Mises criterion, $2k = 1.155Y$, so $P_{\perp} = 2.97Y$. This plane-strain indentation is analogous to a two-dimensional hardness test. A frequently used rule of thumb is that with consistent units, the hardness is approximately three times the yield strength.

HODOGRAPHS

Hodographs are plots of velocity fields. Construction of a hodograph for a slip-line field requires that the field is kinematically admissible. Hodographs can be used to determine where in the field most of the energy is expended and to predict distortion of material as it passes through the field.

In constructing hodographs, it may be noted that:

1. The velocity is constant within a constant pressure zone.
2. In leaving a field of changing pressure there may, or may not be, a sudden change of velocity.
3. The magnitude of the velocity everywhere along a given slip line is the same though the direction may change.
4. In a field of curved lines, the both the magnitude and direction of the velocity change.
5. There is always shear on the boundary between the deforming material inside the field and the rigid material outside of it.
6. The vector representing a velocity discontinuity must be parallel to the discontinuity itself.

Figure 6.10 shows half of the field in Figure 6.8 and the corresponding hodograph. Region OAD moves downward with the velocity, V_o, of the punch. There is a discontinuity, V^*_{OA}, along OA such that the absolute velocity is parallel to the arc AE at point A, and the velocity just to the right of OA differs from that in triangle OAD by a vector parallel to OA. The discontinuity, V^*_A, between the material in the field at A and outside the field equal to the absolute velocity inside the field at A.

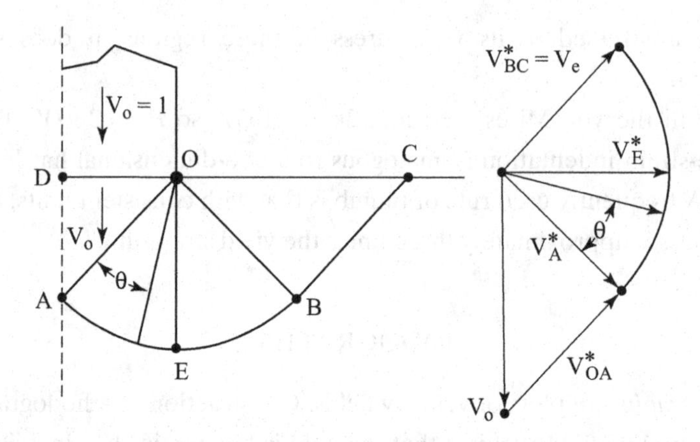

Figure 6.10. (a) A partial slip-line field for indentation and (b) the corresponding hodograph. From W. F. Hosford and R. M. Caddell, *Metal Forming: Mechanics and Metallurgy*, 4th ed. Cambridge University Press (2011).

This discontinuity between material inside and outside the field has a constant magnitude but changing direction along the arc AEB. There is no abrupt velocity discontinuity along OB, so the velocity in triangle OBC is parallel to BC.

In this field, there is intense shear along OA (V^*_{OA}), along AEB ($V^*_A = V^*_E = V^*_B$) and BC ($V^*_{BC} = V^*_A$). Energy is also dissipated by the gradual deformation in the fan OAB.

INDENTATION OF THICK SLABS

Figure 6.11 illustrates the plane-strain indentation of a thick slab by two opposing indenters. The simple slip-line field in Figure 6.12 is appropriate for the special case where the slab thickness, H, equals the indenter width, L. Along AO, $\sigma_x = 0$ so $\sigma_2 = -k$. The stress is the same everywhere in triangle O′OA so along OO′, $\sigma_2 = -k$, $P_\perp = 2k$.

$$P_\perp/2k = 1. \qquad 6.11$$

For larger values of H/L, different fields must be used. Figure 6.13 shows the field for $H/L = 5.43$. This is a field determined by two

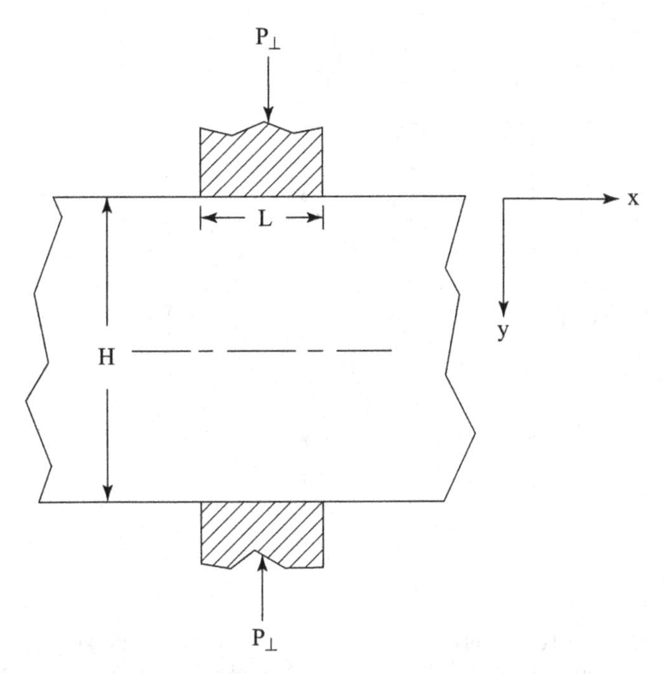

Figure 6.11. Plane-strain indentation of a thick slab by two opposing indenters. From W. F. Hosford and R. M. Caddell, *Metal Forming: Mechanics and Metallurgy*, 4th ed. Cambridge University Press (2011).

centered fans. In triangle O′OA, $\sigma_y = -P_\perp, \sigma_{2(OA)} = \sigma_y = 2 + k = -P_\perp + k$. Moving along an α line to $(0, 1)$, $\sigma_{2(O,1)} = -\sigma_{2(OA)} + 2k\Delta\phi_\alpha$ and moving back along a β line to $(1, 1)$, $\sigma_{2(1,1)} = -\sigma_{2(OA)} + 2k(\Delta\phi_\alpha - \Delta\phi_\beta)$. At point $1, 1, \sigma_{x(1,1)} = P_\perp - +2k + 2k(\Delta\phi_\alpha - \Delta\phi_\beta)$ and at every point (n, n) along the centerline $\sigma_x(n, n) = -P_\perp + 2k + 2k(\Delta\phi_\alpha - \Delta\phi_\beta)_n$. Since $\Delta\phi_\alpha = -\Delta\phi_\beta$,

$$\sigma_{x(n,n)} = -P_\perp + 2k + 2k\Delta\phi_n, \qquad 6.12$$

where $\Delta\phi_n$ is the absolute value of the rotations.

Because there is no constraint in the x-direction,

$$F_x = 0 = \int_0^{H/2} \sigma_x dy. \qquad 6.13$$

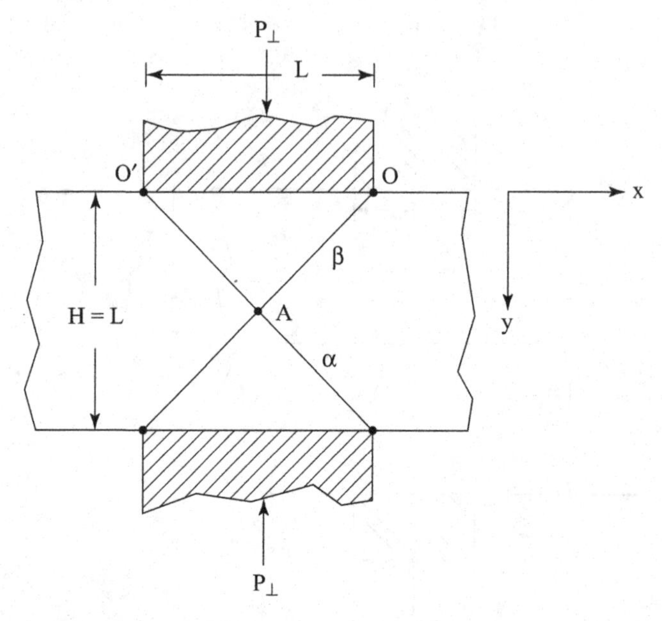

Figure 6.12. Special case of plane-strain indentation where the slab thickness H, equals the indenter width, L. From W. F. Hosford and R. M. Caddell, *Metal Forming: Mechanics and Metallurgy*, 4th ed. Cambridge University Press (2011).

Substituting equation 6.12 into equation 6.13, $\int_0^{H/2}[-P_\perp = 2k(1+2\Delta\phi)]dy = 0$ or

$$P_\perp = 2k + (4k/H)\int_0^{H/2} 2\Delta\phi dy. \qquad 6.14$$

Figure 6.14 gives the values of $\Delta\phi$ are given as a function of y for nodal points on a 15 degree net. There is a more detailed net in the Appendix at the end of this chapter. The integration in equation 6.14 can be done either numerically using the trapezoidal rule or graphically by plotting $2\,\Delta\phi$ versus y. The results of such calculations are summarized in a plot of $P_\perp/2k$ versus H/L (Figure 6.15). It should be noted that for $H/L > 8.75$, $P_\perp/2k$ exceeds $1+\pi/2$ so the field in Figure 6.8 gives a lower value of $P_\perp$, $P_\perp/2k = 1+\pi/2$. Such non-penetrating indentation should be expected for $H/L > 8.75$ and penetrating deformation for $H/L < 8.75$. A corollary is that for valid hardness testing the thickness of the

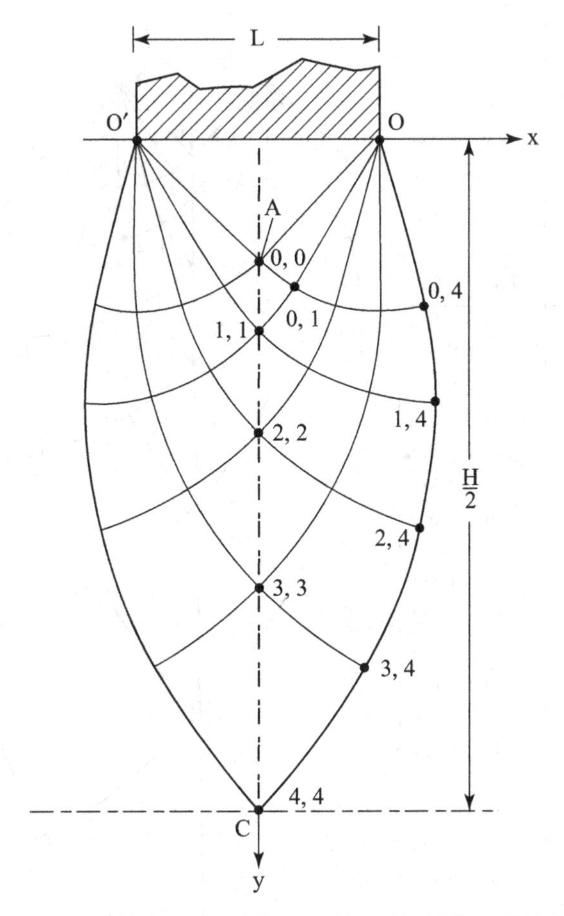

Figure 6.13. Slip-line field for mutual indentation with $H/L = 5.43$. From W. F. Hosford and R. M. Caddell, *Metal Forming: Mechanics and Metallurgy*, 4th ed. Cambridge University Press (2011).

material should be four to five times as thick as the diameter of the indenter. (Theoretically for an indentation on a frictionless substrate $H/L < 8.75/2 = 4.37$.)

CONSTANT SHEAR STRESS INTERFACES

Slip-line fields can be used to solve problems with sticking friction or with a constant friction interface. One example is the compression

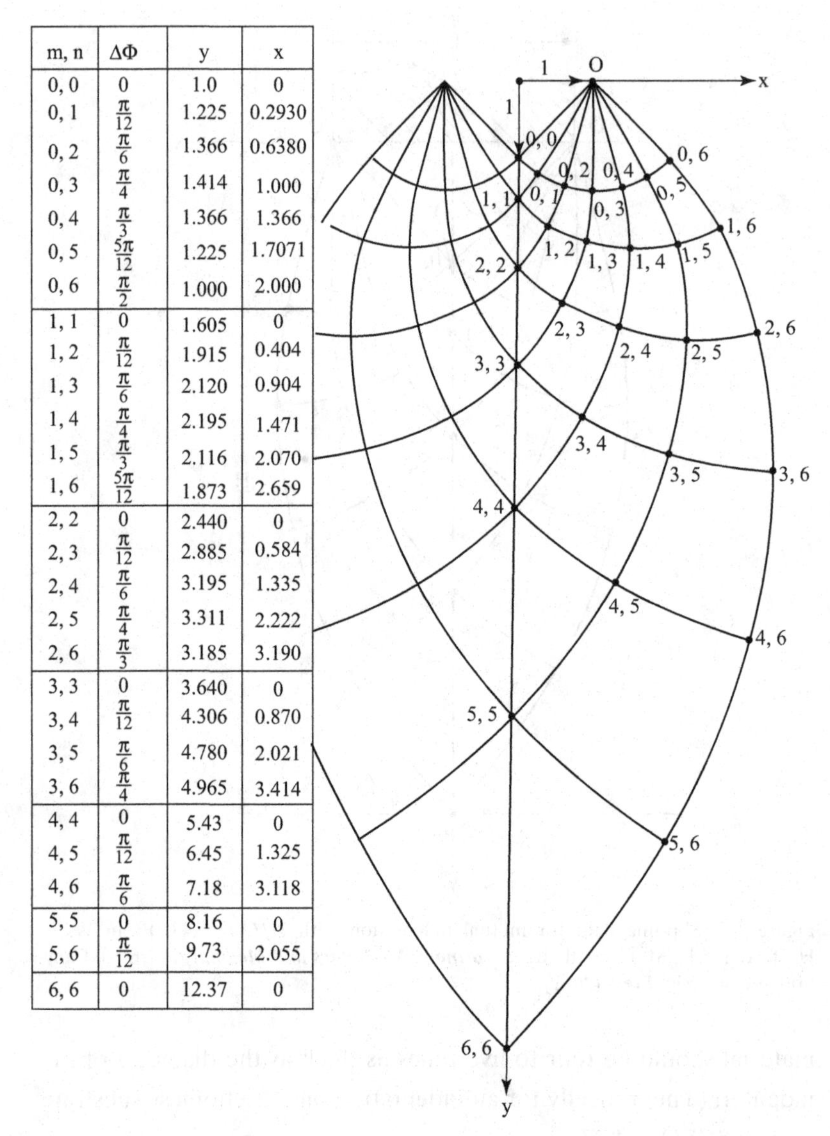

m, n	$\Delta\Phi$	y	x
0, 0	0	1.0	0
0, 1	$\frac{\pi}{12}$	1.225	0.2930
0, 2	$\frac{\pi}{6}$	1.366	0.6380
0, 3	$\frac{\pi}{4}$	1.414	1.000
0, 4	$\frac{\pi}{3}$	1.366	1.366
0, 5	$\frac{5\pi}{12}$	1.225	1.7071
0, 6	$\frac{\pi}{2}$	1.000	2.000
1, 1	0	1.605	0
1, 2	$\frac{\pi}{12}$	1.915	0.404
1, 3	$\frac{\pi}{6}$	2.120	0.904
1, 4	$\frac{\pi}{4}$	2.195	1.471
1, 5	$\frac{\pi}{3}$	2.116	2.070
1, 6	$\frac{5\pi}{12}$	1.873	2.659
2, 2	0	2.440	0
2, 3	$\frac{\pi}{12}$	2.885	0.584
2, 4	$\frac{\pi}{6}$	3.195	1.335
2, 5	$\frac{\pi}{4}$	3.311	2.222
2, 6	$\frac{\pi}{3}$	3.185	3.190
3, 3	0	3.640	0
3, 4	$\frac{\pi}{12}$	4.306	0.870
3, 5	$\frac{\pi}{6}$	4.780	2.021
3, 6	$\frac{\pi}{4}$	4.965	3.414
4, 4	0	5.43	0
4, 5	$\frac{\pi}{12}$	6.45	1.325
4, 6	$\frac{\pi}{6}$	7.18	3.118
5, 5	0	8.16	0
5, 6	$\frac{\pi}{12}$	9.73	2.055
6, 6	0	12.37	0

Figure 6.14. Slip-line field for two centered fans with x and y values for each node of the 15° net. From W. F. Hosford and R. M. Caddell, *Metal Forming: Mechanics and Metallurgy*, 4th ed. Cambridge University Press (2011).

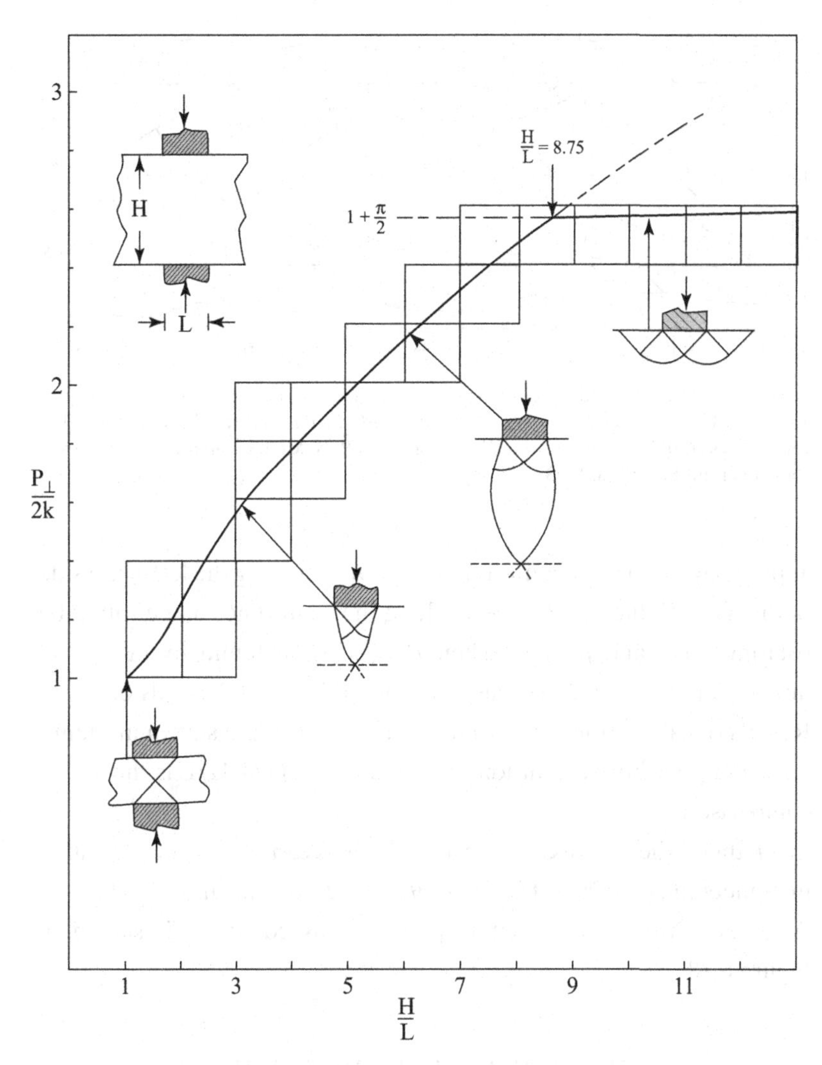

Figure 6.15. Normalized indentation pressure as a function of H/L. This type of plot was first suggested by R. Hill [2]. From W. F. Hosford and R. M. Caddell, *Metal Forming: Mechanics and Metallurgy*, 4th ed. Cambridge University Press (2011).

between rough platens. This approximates conditions during hot forging. With sticking friction the slip lines are parallel and perpendicular to the platens. There is a dead metal zone where they do not meet the platens. Figure 6.16 shows the appropriate slip-line field. The

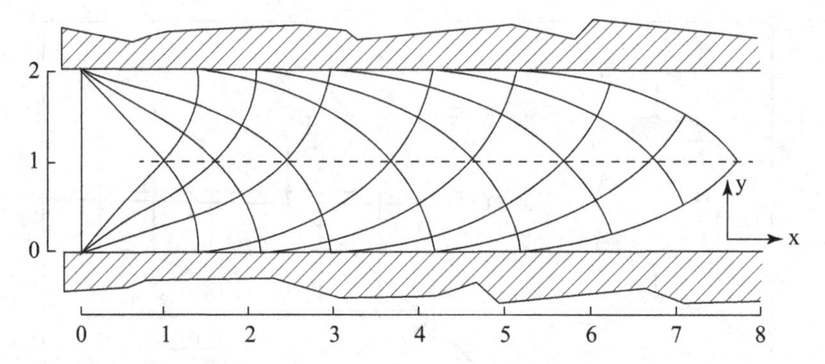

Figure 6.16. Slip-line field for compression with sticking friction. From W. F. Hosford and R. M. Caddell, *Metal Forming: Mechanics and Metallurgy*, 4th ed. Cambridge University Press (2011) Adapted from W. Johnson and P. B. Mellor, *Engineering Plasticity*, Van Nostrand Reinhold, 1973.

appropriate boundary condition is $\sigma_x = \sigma_1 = 0$ along the left-hand side of the field. Values of $P_\perp = -\sigma_x$ along the centerline can be found by rotating on α and β lines. Then $P_{\perp\text{ave}}$ can be found by numerical integration. How much of this field should be used depends on H/L. Results of calculations for various values of H/L are shown in Figure 6.17, an upper-bound solution of $P_\perp/2\text{k} = 1 + (1/4)\text{H/L}$ is shown for comparison.

If there shear stress at the tool interface, $\tau = mk$, the α and β lines meet the interface at an angle, $\theta = (1/2)\cos^{-1}m$ and $\theta' = 90 - \theta$. A general Mohr's stress circle plot for this condition is shown in Figure 6.18.

EXPERIMENTAL EVIDENCE

In 1860, Lüders [1] first noted the networks of orthogonal lines that appear on soft cast steel specimens after bending and etching in nitric acid. These correspond to slip lines. An example of slip lines revealed by etching is given in Figure 6.19. Figure 6.20 shows other examples of slip lines on deformed parts.

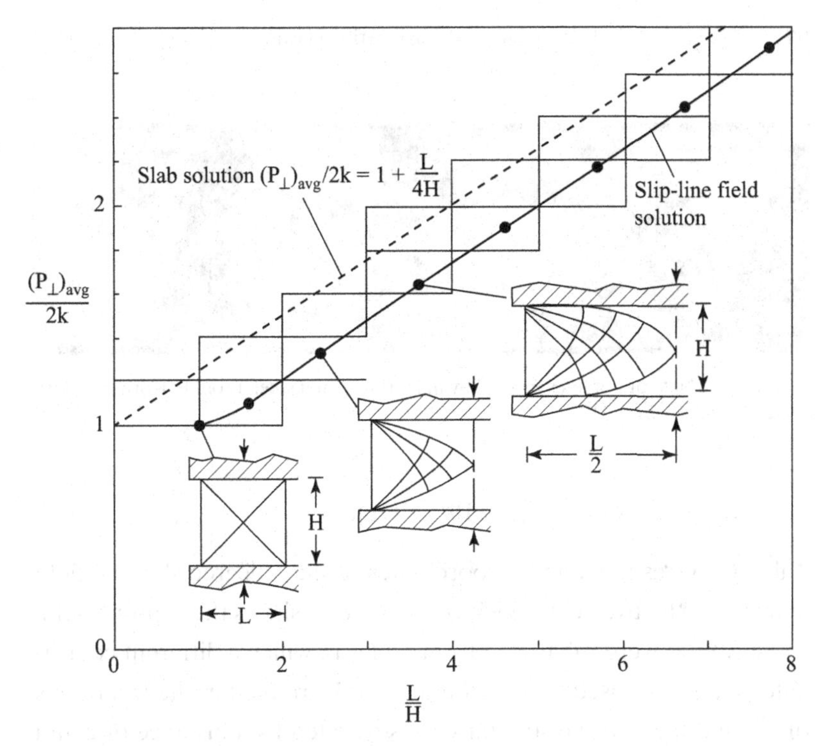

Figure 6.17. Average indentation pressure for the slip-line fields in Fig. 3.14 and an upper-bound solution. From W. F. Hosford and R. M. Caddell, *Metal Forming: Mechanics and Metallurgy*, 4th ed. Cambridge University Press (2011).

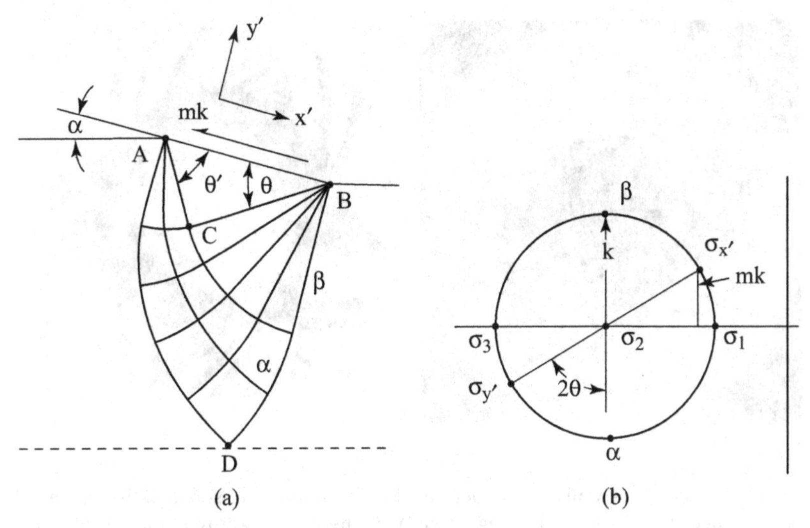

Figure 6.18. (a) Slip-line field for interface stress, $\tau = mk$, and (b) the corresponding Mohr's stress circle diagram. From W. F. Hosford and R. M. Caddell, *Metal Forming: Mechanics and Metallurgy*, 4th ed. Cambridge University Press (2011).

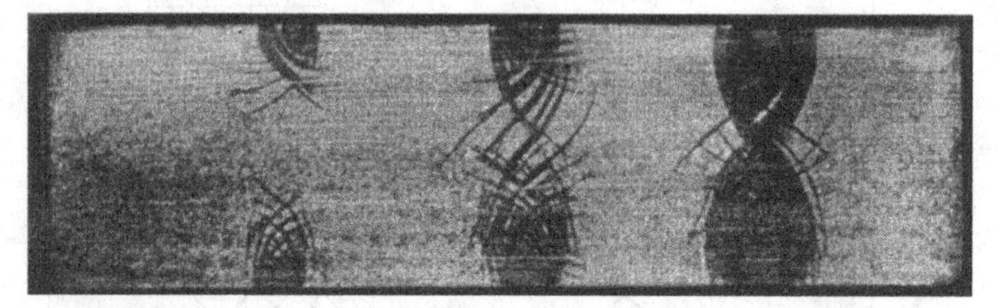

Figure 6.19. Network of lines formed by indenting a mild steel. From F. Körber, *J. Inst. Metals* v. 48. (1932) p. 317.

APPENDIX

Table 6.1 gives the x and y coordinates of the 5 degree slip-line field determined by two centered fans. Figure 6.20 shows the slip-line field. These values were calculated from tables in which a different coordinate system was used to describe the net. Here the fans have a radius of $\sqrt{2}$ and the nodes of the fans are separated by a distance of 2 and the origin is halfway between the nodes.

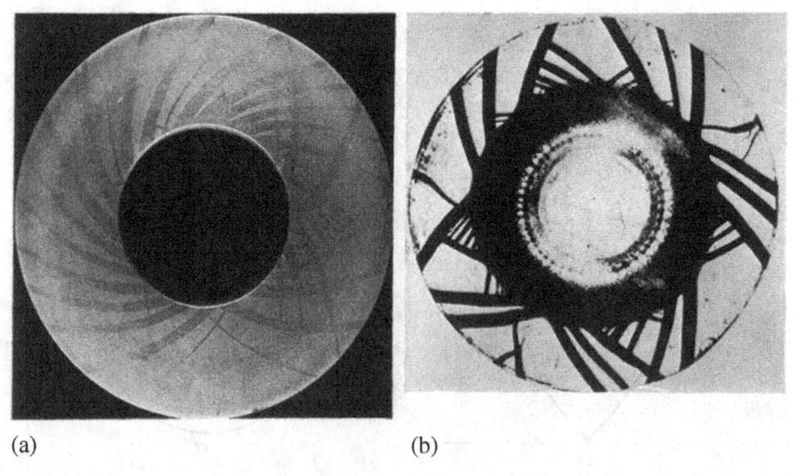

(a) (b)

Figure 6.20. (a) Thick wall cylinder deformed under internal pressure and (b) slip lines on the flange of a cup during drawing. From W. Johnson, R. Sowerby and J. B. Haddow, *Plane-Strain Slip Line Fields*, American Elsevier, 1970.

Table 6.1. *Coordinates of a 5 degree net for the slip-line field determined by two centered fans*

$\Delta\phi_\alpha$	n	m = n	m = n + 1	m = n + 2	m = n + 3	m = n + 4	m = n + 5	m = n + 6	m = n + 7	m = n + 8
0°	0	y = 1.0	1.0833	1.1584	1.2247	1.2817	1.3288	1.3660	1.3926	1.4087
		x = 0	0.0910	0.1888	0.2929	0.4023	0.5163	0.6340	0.7544	0.8767
5°	1	1.1826	1.2741	1.3572	1.4312	1.4951	1.5484	1.5907	1.6214	1.6399
		0.0	0.1000	0.2083	0.3243	0.4472	0.5762	0.7101	0.8484	0.9897
10°	2	1.3831	1.4845	1.5770	1.6597	1.7320	1.7925	1.8407	1.8760	1.8975
		0.0	0.1106	0.2312	0.3613	0.4999	0.6463	0.7995	0.9583	1.1218
15°	3	1.6050	1.7177	1.8214	1.9146	1.9963	2.0653	2.1206	2.1611	2.1861
		0.0	0.1232	0.2582	0.4046	0.5617	0.7285	0.9038	1.0868	1.2760
20°	4	1.8519	1.9781	2.0946	2.2001	2.2929	2.3718	2.4351	2.4820	2.5108
		0.0	0.1377	0.2898	0.4554	0.6339	0.8243	1.0257	1.2307	1.4562
25°	5	2.1283	2.2701	2.4018	2.5215	2.6272	2.7176	2.7905	2.8446	2.8781
		0.0	0.1550	0.3267	0.5146	0.7181	0.9364	1.1680	1.4118	1.6665
30°	6	2.4390	2.5991	2.7484	2.8846	3.0056	3.1093	3.1934	3.2560	3.2948
		0.0	0.1749	0.3698	0.5839	0.8166	1.0610	1.3340	1.6162	1.9119
35°	7	2.7897	2.9713	3.1413	3.2968	3.4356	3.5549	3.6519	3.7245	3.7696
		0.0	0.1984	0.4200	0.6647	0.9314	1.2196	1.5278	1.8547	2.1985
40°	8	3.1874	3.3940	3.5879	3.7662	3.9257	4.0632	4.1755	4.2595	4.3121
		0.0	0.2257	0.4787	0.7589	1.0655	1.3977	1.7540	2.1332	2.5331
45°	9	3.6394	3.8755	4.0976	4.3023	1.1859	4.6447	4.7747	4.8732	4.9335
		0.0	0.2575	0.5472	0.8688	1.2219	1.6054	2.0182	2.4586	2.9243
50°	10	4.1561	4.4259	4.6808	4.9162	5.1281	5.3117	5.4626	5.5760	5.6472
		0.0	0.2947	0.6272	0.9973	1.4046	1.8482	2.3269	2.8389	3.3828
55°	11	4.7470	5.0565	5.3496	5.6211	5.8659	6.0786	6.2537	6.3856	
		0.0	0.3380	07205	1.1472	1.6179	2.1318	2.6873	3.2831	
60°	12	5.4248	5.7807	6.1185	6.4321	6.7154	6.9622	7.1657		
		0.0	0.3886	0.8296	1.3223	1.8670	2.4631	3.1091		
65°	13	6.2043	6.6144	7.0043	7.3671	7.6955	7.9820			
		0.0	0.4982	0.9573	1.5269	2.1584	2.8505			
70°	14	7.1023	7.5758	8.0267	8.4470	8.8281				
		0.0	0.5172	1.1055	1.7658	2.4986				
75°	15	8.1290	8.6864	9.2085	9.6961					
		0.0	0.5981	1.2794	2.0455					
80°	16	9.3375	9.9715	10.5771						
		0.0	0.6925	1.4827						
85°	17	10.726	11.460							
		0.0	0.8031							
90°	20	12.334								
		0.0								

(*continued*)

Table 6.1 *(continued)*

$\Delta\phi_\alpha$	n	n + 9	n + 10	n + 11	n + 12	n + 13	n = 14	n + 15	n + 16	n + 17	n + 18
0°	0	1.4141	1.4087	1.3926	1.3629	1.3288	1.2816	1.2246	1.1584	1.0833	1.000
		1.0000	1.1233	1.2456	1.3660	1.4837	1.5977	1.7071	1.8112	1.9090	2.000
5°	1	1.6463	1.6399	1.6068	1.5892	1.5449	1.4879	1.4189	1.3379	1.2455	
		1.1334	1.2718	1.4222	1.5653	1.7061	1.8434	1.9765	2.1036	2.2240	
10°	2	1.9048	1.8975	1.8751	1.8375	1.7846	1.7163	1.6330	1.5347		
		1.2891	1.4582	1.6282	1.7979	1.9658	2.1305	2.2909	2.4452		
15°	3	2.1946	2.1860	2.1595	2.1151	2.0522	1.9707	1.8707			
		1.4708	1.6686	1.8688	2.0694	2.2690	2.4658	2.6583			
20°	4	2.5207	2.5107	2.4797	2.4272	2.3527	2.2555				
		1.6828	1.9141	2.1497	2.3865	2.6233	2.8574				
25°	5	2.8895	2.8778	2.8414	2.7795	2.6913					
		1.9304	2.2014	2.4777	2.7580	3.0372					
30°	6	3.3083	3.2944	3.2519	3.1789						
		2.2196	2.366	2.8610	3.1901						
35°	7	3.7853	3.7692	3.7191							
		2.5573	2.9281	3.3089							
40°	8	4.3303	4.3114								
		2.9518	3.3859								
45°	9	4.9548									
		3.4129									

NOTE OF INTEREST

Prandl [2] was the first to utilize slip-line field theory in analysis of indentation of semi-infinite blocks. The concept of hodgraphs was introduced by Prager [3] and applied to plane-strain problems by Green [4]. Equations 5.4 were first proposed in (1921) by Hencky [5]. The concept of hodgraphs was introduced by Prager [9] and applied to plane-strain problems by Green. Several books [7, 8] review the development of slip-line fields.

W. Johnson did much to show how slip-line fields could be used to analyze metal forming operations. He was born in Manchester in 1922 then went on to graduate from UMIST (University of Manchester) in 1943. After graduating in engineering from UMIST, he served with the Royal Electrical and Mechanical Engineers in Italy. After the war, he entered the civil service but in 1950 he took up a lectureship at Northampton Institute. In 1952, he moved to Sheffield University, where he did research metal forming. At 38, he was appointed to the chair of mechanical engineering at the University of Manchester Institute of Science and Technology (UMIST), where he won a reputation for research and teaching. Johnson established the *International Journal of Mechanical Science* and the *International Journal of Impact Engineering*. In 1975, he took up the chair of mechanics at Cambridge University. He was elected to the Royal Society and to the Royal Academy of Engineering, and received honorary degrees from UMIST and the Universities of Sheffield and Bradford. He retired from British university life in 1983 to spend three years at Purdue University in Indiana as professor. He died at age 88 having published more than 400 research papers and eight books in his field of applied mechanics.

REFERENCES

1. W. Lüders, *Dinglers Polytech., J.* Stuttgart, 1860.
2. L. Prandl, *Zeits. Angew. Math. Mech.* v. 1. (1921).
3. H. Hencky, *Zeits. Angew. Math. Mech.* v. 1. (1921).
4. W. Prager, *Trans. R. Inst. Tech., Stockholm*, no. 5 (1953).
5. A. P. Green, *J. Mech. Phys. Solids* v.2 (1974).
6. W. Johnson, R. Sowrby and J. B. Haddow, *Plain-strain Slip-line Fields: Theory and Bibliography*, American Elsevier (1970).
7. W. Johnson, R. Sowerby and R. D. Venter, *Plane-Strain Slip-line Fields for Metal Deformation Processes*, Pergamon Press (1982).

ADDITIONAL REFERENCES

1. W. F. Hosford and R. M. Caddell, *Metal Forming: Mechanics and Metallurgy*, 4th ed., Cambridge University Press (2011).
2. R. Hill, *The Mathematical Theory of Plasticity*, Oxford University Press (1950).

7

ANISOTROPIC PLASTICITY

Although materials are frequently assumed to be isotropic, they rarely are. The two main causes of anisotropy are *preferred orientations* of grains (*crystallographic texture*) and *mechanical fibering*, which reflect the elongation and alignment of microstructural features such as inclusions and grain boundaries. Anisotropy of plastic behavior is almost entirely caused by the presence of preferred orientations.

HILL'S 1948 ANISOTROPIC THEORY

In 1948, Hill [1, 2] proposed the first quantitative treatment of plastic anisotropy. He considered materials with three orthogonal axes of anisotropy, x, y, and z about which there is a two-fold symmetry of properties. Therefore, the yz, zx, and xy planes are planes of mirror symmetry. In a rolled sheet, the x, y, and z-axes are usually taken as the rolling direction, the transverse direction and the sheet-plane normal.

Hill's yield criterion is a generalization of the von Mises criterion:

$$F(\sigma_y - \sigma_z)^2 + G(\sigma_z - \sigma_x)^2 + H(\sigma_x - \sigma_y)^2 + 2L\tau_{yz}^2 + 2M\tau_{zx}^2 + 2N\tau_{xy}^2 = 2f(\sigma_{ij})^2, \quad 7.1$$

where *F, G, H, L, M*, and *N* are constants that describe the anisotropy. Note that if $F = G = H = 1$ and $L = M = N = 3$, this reduces to the von Mises criterion. The constants *F, G* and *H* can be evaluated from simple tension tests. If the x-direction yield strength is X, a yielding in an x-direction tension test, $\sigma_x = X$ and $\sigma_y = \sigma_z = \tau_{ij} = 0$. Substituting into equation 6.1, $(G+H)X^2 = 1$ *or* $X^2 = 1/(G+H)$. Similarly,

$$X^2 = \frac{1}{G+H}, \quad Y^2 = \frac{1}{H+F} \quad \text{and} \quad Z^2 = \frac{1}{F+G}. \qquad 7.2$$

Solving simultaneously,

$$\begin{aligned} F &= (1/Y^2 + 1/Z^2 - 1/X^2)/2 \\ G &= (1/Z^2 + 1/X^2 - 1/Y^2)/2, \\ H &= (1/X^2 + 1/Y^2 - 1/Z^2)/2 \end{aligned} \qquad 7.3$$

where Y and Z are the y- and z-direction yield strengths. Measurement of Z, however, is not feasible for sheets.

Using the general flow rule, (equation 4.20),

$$d\varepsilon_{ij} = d\lambda \frac{\partial f(\sigma_{ij})}{\partial \sigma_{ij}}, \qquad 7.4$$

the flow rules become:

$$\begin{aligned} d\varepsilon_x &= d\lambda[H(\sigma_x - \sigma_y) + G(\sigma_x - \sigma_z)], \quad d\varepsilon_{yz} = d\varepsilon_{zy} = d\lambda L\tau_{zy} \\ d\varepsilon_y &= d\lambda[F(\sigma_y - \sigma_z) + H(\sigma_y - \sigma_x)], \quad d\varepsilon_{zx} = d\varepsilon_{zx} = d\lambda M\tau_{zx} \\ d\varepsilon_z &= d\lambda[G(\sigma_z - \sigma_x) + F(\sigma_z - \sigma_y)], \quad d\varepsilon_{xy} = d\varepsilon_{yx} = d\lambda N\tau_{xy} \end{aligned} \qquad 7.5$$

To derive these flow rules, the yield criterion must be written with the shear stress terms appearing as $L(\tau_{yz}^2 + \tau_{zy}^2) + M(\tau_{zx}^2 + \tau_{xz}^2) + N(\tau_{xy}^2 + \tau_{yx}^2)$. Note that in equation 7.5, $d\varepsilon_x + d\varepsilon_y + d\varepsilon_z = 0$ indicates constant volume.

For an x-direction tension test at yielding, $\sigma_x = X, \sigma_y = \sigma_z = 0$. Substituting into equation 7.5, $d\varepsilon_x = d\lambda(H+G)X$, $d\varepsilon_y = d\lambda HX$ and $d\varepsilon_x = d\lambda GX$. The strain ratio, R, in an x-direction tension test is defined as $d\varepsilon_y/d\varepsilon_z$ so

$$R = H/G. \qquad 7.6$$

Defining P as $d\varepsilon_x/d\varepsilon_z$ in a y-direction tension test

$$P = H/F. \qquad 7.7$$

With equations 7.6 and 7.7, the yield strength, Z, can be calculated by measuring, *X, Y, R*, and *P* in x- and y- direction tests.

$$Z^2/X^2 = (G+H)/(F+G) = [(1/R)+1]/1/R+1/P \qquad 7.8$$

or

$$Z = X\sqrt{[P(R+1)/(P+R)]} \qquad 7.9$$

and

$$Z = Y\sqrt{[R(P+1)/(P+R)]} \qquad 7.10$$

The constant, N, in equation 7.1, can be found from a tension test made at an angle, θ, to the x-axis. In such a test, yielding occurs when

$$F\sin^4\theta + G\cos^4\theta + H(\cos^4\theta - 2\cos^2\theta\sin^2\theta + \sin^4\theta) + 2N\cos^2\theta\sin^2\theta = Y_\theta^2, \qquad 7.11$$

where is the yield strength in the θ–direction test. Equation 7.11 can be simplified to

$$Y_\theta = [H + F\sin^2\theta + G\cos^2\theta + (2N - F - 4H)\cos 2\theta\sin^2\theta)]^{-1/2}. \qquad 7.12$$

For a 45 degree test, this becomes $F/2 + G/2 + N = Y_\theta^2$. Solving for *N*,

$$N = 2Y_{45}^2 - (F+G)/2. \qquad 7.13$$

Differentiating equation 7.11 with respect to θ, it can be shown that there are minima and maxima in the value of Y_θ at $\theta^* = 0°$ and $90°$ and at

$$\theta^* = \arctan[(N - G - 2H]/(N - F - 2H)\}^{-1/2}. \qquad 7.14$$

There are four possibilities:

With $N > G + 2H$ and $N > F + 2H$ maxima occur at 0 degrees and 90 degrees and there is a minimum at θ^*,

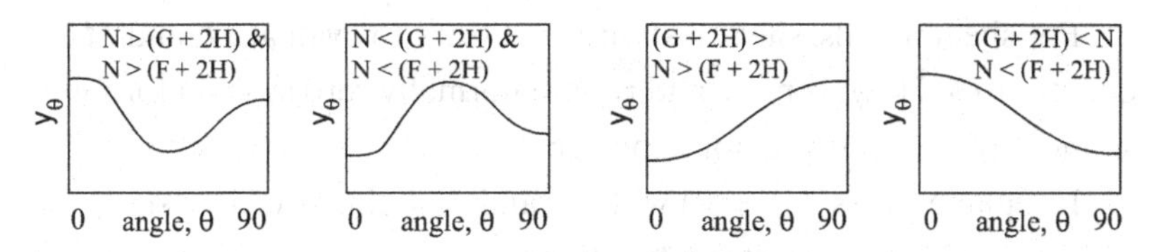

Figure 7.1. Possible variations of yield strength with orientation. From W. F. Hosford, *The Mechanics of Crystals and Textured Polycrystals*, Oxford University Press (1993).

With $N < G + 2H$ and $N < F + 2H$ minima occur at 0 degrees and 90 degrees, and there is a maximum at θ^*,

If $G + 2H > N > F + 2H$, θ^* is imaginary and there is a maximum at 0 degrees and a minimum at 90 degrees,

If $G + 2H < N < F + 2H$, θ^* is imaginary and there is a minimum at 0 degrees and a maximum at 90 degrees.

These possibilities are illustrated in Figure 7.1.

An alternative way of evaluating N is from the R-values. From the flow rules,

$$\begin{aligned} \varepsilon_x &= \lambda(\sigma_{x'}[(H+G)\cos^2\theta - H\sin^2\theta] \\ \varepsilon_y &= \lambda(\sigma_{x'}[(F+H)\cos^2\theta - H\cos^2\theta] \\ \varepsilon_z &= \lambda(\sigma_{x'}[-F\cos^2\theta - G\sin^2\theta] \\ \gamma_{xy} &= \lambda(\sigma_{x'} 2N\cos\theta\sin^2\theta. \end{aligned} \tag{7.15}$$

The strain ratio, R_θ, expressed in terms of the anisotropic parameters becomes

$$R_\theta = [H + (2N - F - 4H)\sin^2\theta\cos^2\theta]/(F\sin^2\theta + G\cos^2\theta).$$

For $\theta = 45°$, $R_{45} = N/(F+G) - 1/2$.

Solving for N, $N = (R_{45} + 1/2)(F+G)$.

Substituting $R = H/G$ and $P = H/F$,

$$N = (2R_{45} + 1)(R+P)/[2(R+1)PX^2]. \tag{7.16}$$

The R-value has a minimum or a maximum at 0 degrees and 90 degrees and there may be one minimum or one maximum between 0 degrees and 90 degrees.

For sheet metals, shear tests are necessary to evaluate L and M in equation 7.1. However, τ_{yz} and τ_{zx} are normally zero in sheet forming so these parameters are not important.

For plane stress, ($\varepsilon_z = 0$) with σ_x and σ_y being principal stresses, the effective stress and strain functions are:

$$\bar{\sigma} = \left[\frac{P|\sigma_x|^2 + R|\sigma_y|2 + RP|\sigma_x - \sigma_y|^2}{P(R+1)}\right]^{1/2} \qquad 7.17$$

and

$$\bar{\varepsilon} = \varepsilon_x(1+\alpha\rho)\frac{\sigma_x}{\bar{\sigma}}, \qquad 7.18$$

where

$$\rho = \frac{d\varepsilon_y}{d\varepsilon_x} = \frac{R[\alpha - P(1-\alpha)]}{P[1+R(1-\alpha)]}.$$

SPECIAL CASES OF HILL'S 1948 YIELD CRITERION

For the special case in which x, y, and z are principal axes ($\tau_{yz} = \tau_{zx} = \tau_{xy} = 0$), equation 7.1 can be expressed in terms of R and $P = R_{90}$. Substituting $(G+H)X^2 = 1$,

$$\left(\frac{F}{G}\right)(\sigma_y - \sigma_z)^2 + \left(\frac{G}{G}\right)(\sigma_z - \sigma_x)^2 + \left(\frac{H}{G}\right)(\sigma_x - \sigma_y)^2 = \left[\left(\frac{G}{G}\right) + \left(\frac{H}{G}\right)\right]X^2. \qquad 7.19$$

Now substituting, $R = H/G$ and $R/P = F/G$ and multiplying by P,

$$R(\sigma_y - \sigma_z)^2 + P(\sigma_z - \sigma_x)^2 + RP(\sigma_x - \sigma_y)^2 = P(R+1)X^2. \quad 7.20$$

The flow rules become

$$\varepsilon_x : \varepsilon_y : \varepsilon_z = R(\sigma_x - \sigma_y) + (\sigma_z - \sigma_z) : (R/P)(\sigma_y - \sigma_z) + R(\sigma_y - \sigma_x) : (R/P)(\sigma_z - \sigma_y) + (\sigma_z - \sigma_x). \qquad 7.21$$

Expressing the effective stress and effective strain for this criterion in a way that they reduce to σ_x and ε_x in a uniaxial tension test [3]:

$$\bar{\sigma} = \left[\sigma_1^2 + \sigma_2^2 + 2R\sigma_1\sigma_2/(R+1)\right]^{1/2} \qquad 7.22$$

and

$$\bar{\varepsilon} = C[P(\varepsilon_y - R\varepsilon_z)^2 + R(P\varepsilon_z - \varepsilon_x)^2 + (R\varepsilon_x - P\varepsilon_y)^2]^{1/2}$$

$$\bar{\varepsilon} = \frac{1+R}{(1+2R)^{1/2}}\left[\varepsilon_1^2 + \varepsilon_2^2 + \varepsilon_1\varepsilon_2\right]^{1/2}. \qquad 7.23$$

With rotational symmetry about the z-direction, $F = G = H = N/3$, $L = M$ and $R = P = R_{45}$. The x and y directions may be chosen to coincide with the principle stress axes so $\tau_{xy} = 0$. Substituting $P = R$ into equation 7.20 and the flow rules, flow rules become:

$$(\sigma_y - \sigma_z)^2 + (\sigma_z - \sigma_x)^2 + R(\sigma_x - \sigma_y)^2 = (R+1)X^2 \qquad 7.24$$

and

$$\varepsilon_x : \varepsilon_y : \varepsilon_z$$
$$= (R+1)\sigma_x - R\sigma_y - \sigma_z : (R+1)\sigma_y - R\sigma_x - \sigma_z : 2\sigma_z - R\sigma_x - \sigma_y. \qquad 7.25$$

Equation 7.24 plots in $\sigma_z = 0$ space as an ellipse as shown in Figure 7.2. The extension of the ellipse into the first quadrant increases with increasing R-value indicating that the strength in biaxial tension increases with R.

The R-value varies with direction. However, to use equations 7.24 and 7.25, it is common to take an average value as

$$\bar{R} = (R_0 + 2R_{45} + R_{90})/4. \qquad 7.26$$

While this procedure is not strictly correct, it is often used to assess the role of normal anisotropy.

For rotational symmetry about z, the effective stress and strain functions reduce to

$$\bar{\sigma} = \left[\sigma_1^2 + \sigma_2^2 - 2R\sigma_1\sigma_2/(1+R)\right]^{1/2}$$

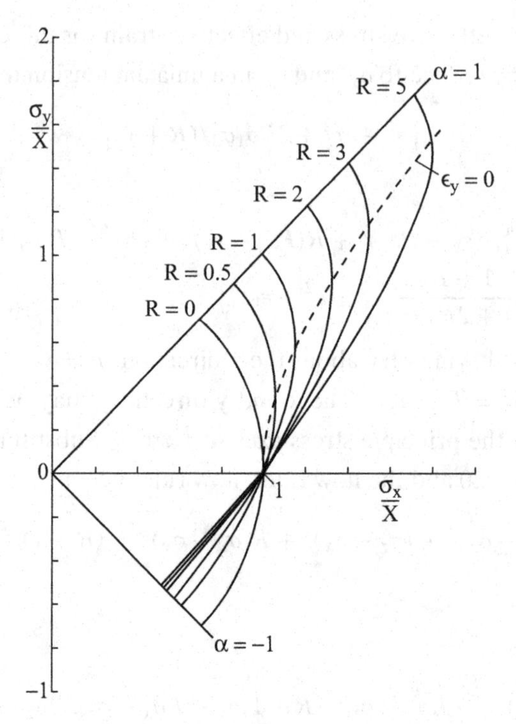

Figure 7.2. Plane-stress ($\sigma_z = 0$) yield locus for rotational symmetry about z according to the Hill criterion (equation 6.18). The dashed line indicates the locus of stress states for plane strain, $\varepsilon_y = 0$. From W. F. Hosford, *The Mechanics of Crystals and Textured Polycrystals*, Oxford University Press (1993).

and

$$\bar{\varepsilon} = \left\{ \frac{1+R}{(1+2R)^{1/2}} \left[\varepsilon_1^2 + \varepsilon_2^2 + 2R\varepsilon_1\varepsilon_2/(1+R)\right] \right\}^{1/2}. \quad 7.27$$

NON-QUADRATIC YIELD CRITERIA

Calculations [4, 5] of yield loci for textured fcc and bcc metals suggested that a non-quadratic yield criterion of the form

$$F|\sigma_y - \sigma_z|^a + G|\sigma_z - \sigma_x|^a + H|\sigma_x - \sigma y|^a = 1, \quad 7.28$$

with an exponent much higher than 2, represents the anisotropy much better. With $a = 2$ this reduces to equation 7.20. Exponents of 8 for fcc metals and 6 for bcc metals were suggested. Although this criterion is a special case of Hill's 1979 criterion [6], it was suggested independently and it is not one that Hill suggested would be useful.

For plane stress conditions, this criterion reduces to

$$P|\sigma_x|^a + R|\sigma_y|^a + RP|\sigma_x - \sigma y|^a = P(R+1)X^a. \qquad 7.29$$

If the exponent is an even integer, the absolute magnitude signs in equation 7.29 are unnecessary. With rotational symmetry about z, $R = P$ and the criterion reduces to

$$\sigma_x^a + \sigma_y^a + R(\sigma_x - \sigma_y)^a = (R+1)Y^a. \qquad 7.30$$

The yield locus for equation 7.30 plots between the Tresca and Hill 1948 criteria (Figure 7.3). As the exponent, a, increases, the criterion approaches Tresca.

The flow rules for equation 7.29 are:

$$\varepsilon_x = \lambda[P\sigma_x^{a-1} + RP(\sigma_x - \sigma_y)^{a-1}],$$

$$\varepsilon_y = \lambda[R\sigma_y^{a-1} + RP(\sigma_y - \sigma_x)^{a-1}], \qquad 7.31$$

$$\varepsilon_z = -\lambda(P\sigma_x^{a-1} + R\sigma_y^{a-1}).$$

The effective stress and strain functions are:

$$\bar{\sigma} = \left[\frac{P|\sigma_x|^a + R|\sigma_y|^a + RP|\sigma_x - \sigma_y|^a}{P(R+1)}\right]^{1/a} \qquad 7.32$$

and

$$\bar{\varepsilon} = \varepsilon_x(1+\alpha\rho)\frac{\sigma_x}{\bar{\sigma}}, \qquad 7.33$$

where

$$\rho = \frac{d\varepsilon_y}{d\varepsilon_x} = \frac{R[\alpha^{a-1} - P(1-\alpha^{a-1})]}{P[1+R(1-\alpha)^{a-1}]}. \qquad 7.34$$

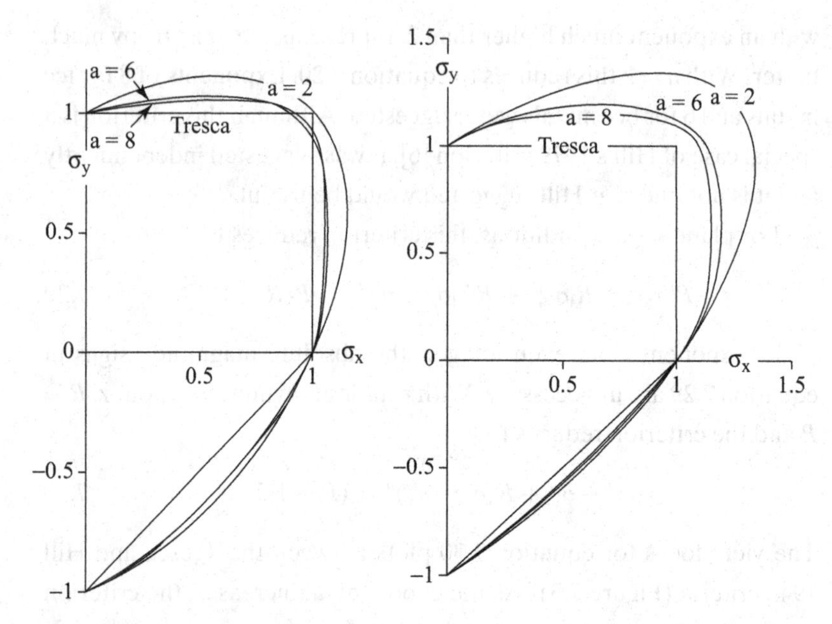

Figure 7.3. Plots of this criterion for several values of $R = 0.5$ (a.) and $R = 2$ (b.) with several values of the exponent *a*. Note as *a* increases, the loci approach the Tresca locus. From W. F. Hosford, *The Mechanics of Crystals and Textured Polycrystals*, Oxford University Press (1993).

Unless $a = 2$, equation 7.34 requires a numerical solution to find the stress ratio, α, from a knowledge of ρ.

In 1979, Hill [6] proposed a generalized non-quadratic criterion to account for an "anomalous" observation [7] that in some aluminum alloys with $R > 1$, $P > 1$ and $R_{45} > 1$, yield strengths in biaxial tension were found to be higher than the yield strength in uniaxial tension. This is not permitted with Hill's 1948 criterion.

$$f|\sigma_2 - \sigma_3|^m + g|\sigma_3 - \sigma_1|^m + h|\sigma_1 - \sigma_2|^m$$
$$+ |\sigma_1 + \sigma_2|^m + (2R+1)|\sigma_1 - \sigma_2|^m = 2(R+1)Y^m$$
$$+ a|2\sigma_1 - \sigma_2 - \sigma_3|^m + b|2\sigma_2 - \sigma_3 - \sigma_1|^m + c|2\sigma_3 - \sigma_1 - \sigma_2|^m = 1, \quad 7.35$$

where the exponent, *m*, depends on the material. Hill suggested four special cases with planar isotropy ($a = b$ *and* $g = h$). Using the

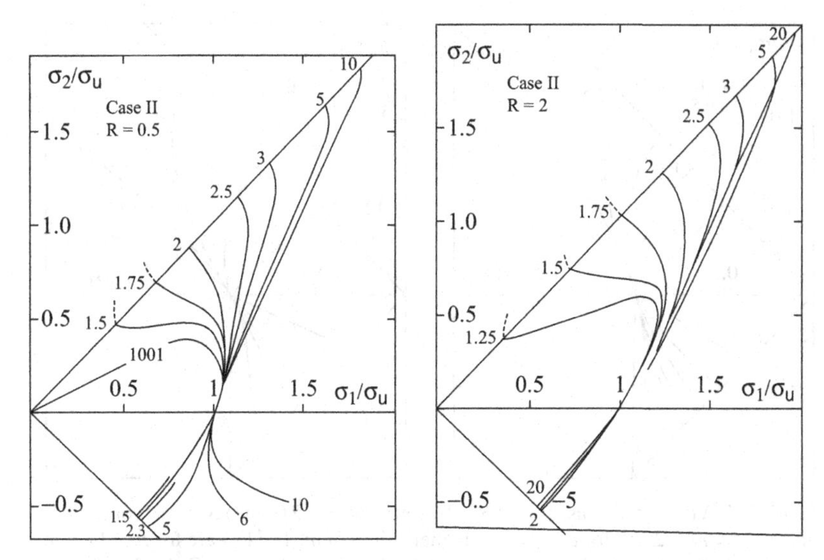

Figure 7.4. Yield loci for case I with $R = 1/2$ (left) and $R = 2$ (right.) Note the concavities for $\alpha = -1/2$ near $m < 2$ and for $\alpha = 0$ for $m > 2$. The loci are unbounded at $\alpha = 1$ for large values of m. From Y. Zhu, B. Dodd, R. M. Caddell and W. F. Hosford, *Int. J. Mech. Sci.*, v. 29 (1987).

corresponding flow rules in these cases to express R,

$$R = \frac{2^{m-1}a + h + 2b - c}{2^{m-1}a + g - b + 2c}. \tag{7.36}$$

Of these, only Hill's fourth case, which can be expressed as

$$|\sigma_1 + \sigma_2|^m + (2R+1)|\sigma_1 - \sigma_2|^m = 2(R+1)Y^m \tag{7.37}$$

is free from concavity problems [8]. Values of $1.7 < m < 2.2$ have been required to fit this to experimental data and different exponents are required for different R-values so this criterion cannot be used to predict the effect of R on forming operations. The other special cases are either outwardly concave or unbounded (Figures 7.4 to 7.7). It should be noted that equation 7.28 is a special case of equation 7.35. However, it does not account for the "anomaly."

The high exponent yield criteria, equations 7.24 and 7.25, do not provide any way of treating shear stress terms, τ_{yz}, τ_{zx}, or τ_{xy}. In 1989,

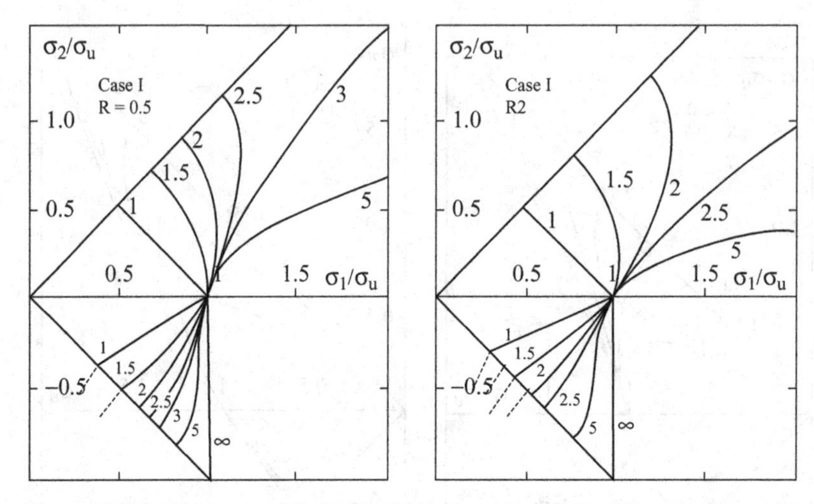

Figure 7.5. Yield loci for case II with $R = 1/2$ (left) and $R = 2$ (right.) Note the concavities at $\alpha < 1$ for $m > 2$ and for $\alpha = 1/2$ for higher values of m. The loci are for $R = 1/2$ and $m = 8$ and 10 are unbounded at $\alpha = 1$. From Y. Zhu, B. Dodd, R. M. Caddell and W. F. Hosford, *Int. J. Mech. Sci.*, v. 29 (1987).

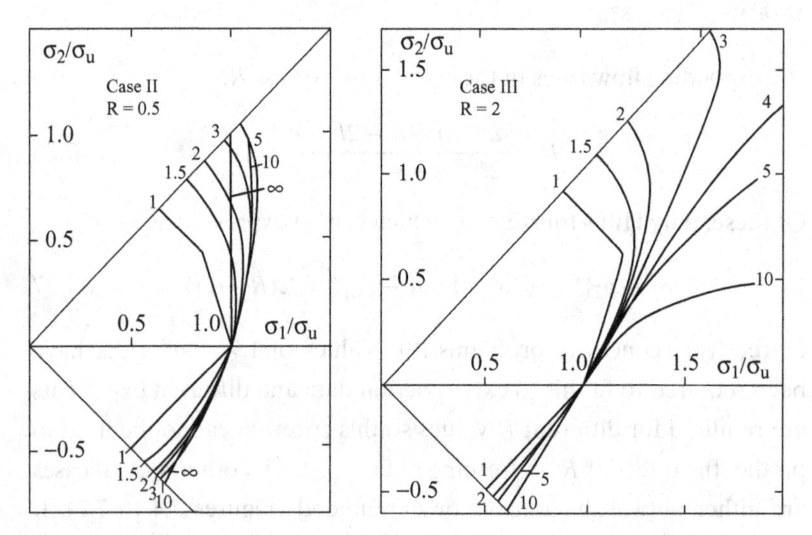

Figure 7.6. Yield loci for case III with $R = 1/2$ (left) and $R = 2$ (right.) There are no concavities for low values of R but for high values there is a concavity for $\alpha = 1/2$. The loci for $R = 2$ and $m > 4$ are unbounded. From Y. Zhu, B. Dodd, R. M. Caddell and W. F. Hosford, *Int. J. Mech. Sci.*, v. 29 (1987).

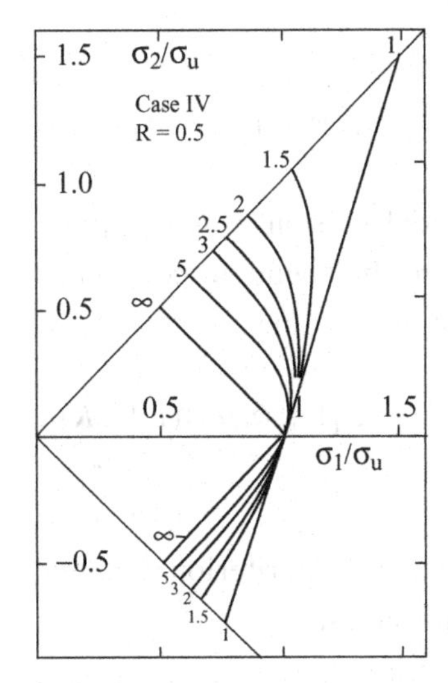

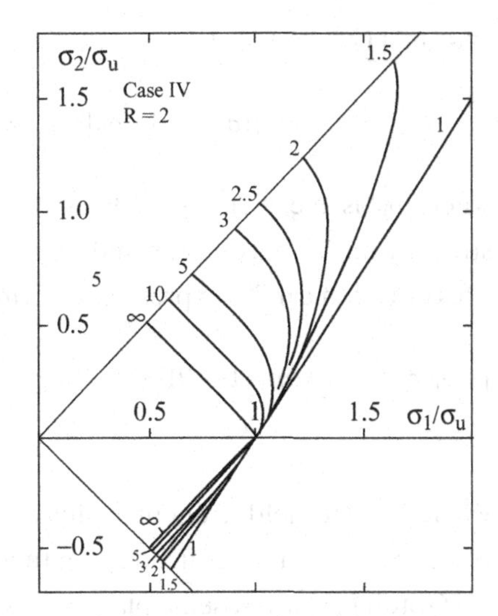

Figure 7.7. Yield loci for case IV with $R = 1/2$ (left) and $R = 2$ (right.) The loci are neither outwardly concave nor unbounded. From Y. Zhu, B. Dodd, R. M. Caddell and W. F. Hosford, *Int. J. Mech. Sci.*, v. 29 (1987).

Barlat and Lian [9] proposed the plane-stress criterion that accounts for the in-plane shear stress, σ_{xy}.

$$a|K_1 - K_2|^m + a|K_1 + K_2|^m + (a-2)|2K_2|^m = 2Y^m \qquad 7.38$$

where $K_1 = (\sigma_x + h\sigma_y)/2$ and $K_2 = \{[(\sigma_x - h\sigma_y)/2]^2 + p\tau_{xy}^2\}^{1/2}$. Here a, p, h, and m are material constants. The exponent, m is approximately 8. It should be noted that this criterion reduces to equation 7.27 for planar isotropy.

Later Barlat and coworkers [10, 11] proposed a criterion that allows for out of plane shear stresses, τ_{yz} and τ_{zx}. However, this criterion requires six constants in addition to m. It will not be discussed further except to state that it does not reduce to equation 7.30 when τ_{yz} and $\tau_{zx} = 0$.

Other workers have proposed anisotropic yield criteria. These generally contain more constants that must be determined experimentally.

Bassani [12] proposed

$$f = |(\sigma_1 + \sigma_2)/\sigma_b|^n + |(\sigma_1 - \sigma_2)/2\tau|^m - 1, \tag{7.39}$$

where σ_b is the yield strength under biaxial tension, τ is the yield strength under pure shear and m and n are independent constants. This criterion can be expressed in terms of the R-value as

$$|\sigma_1 + \sigma_2|^n + (n/m)(1 + 2R)X^{n-m}|\sigma_1 - \sigma_2|^m = X^n[1 + (n/m)(1 + 2R)], \tag{7.40}$$

where X is the yield strength in uniaxial tension. This criterion requires two experimental tests in addition to a tension test.

Gotoh [13] proposed a plane stress criterion

$$f = A_0(\sigma_x + \sigma_y)^2 + A_1\sigma_x^4 + A_2\sigma_x^3\sigma_y + A_3\sigma_x^2\sigma_y^2 + +A_4\sigma_x\sigma_y^3 + +A_5\sigma_y^4 \\ +\left(A_6\sigma_x^2 + A_7\sigma_x\sigma_y + A_8\sigma_y^2\right)\tau^2 + A_9)\tau^4, \tag{7.41}$$

It has ten independent constants. The first term allows for compressibility, so for incompressible materials $A_0 = 0$.

In an attempt to accommodate planar anisotropy, it has been suggested [13] that the yield criterion be expressed in terms of the principal stresses rather than the stress components along the symmetry axes. Modifying equation 7.23,

$$R_\theta\sigma_\theta^a + R_{\theta+90}\sigma_{\theta+90}^a + R_\theta R_{\theta+90}\left(\sigma_\theta^a + \sigma_{\theta+90}\right)^a = R_{\theta+90}(R_\theta + 1)X_\theta^a, \tag{7.42}$$

where σ_θ and $\sigma_{\theta+90}$ and R_θ and $R_{\theta+90}$ are yield strengths and the strain ratios measured in tension tests along the θ and $\theta + 90$ directions. For this criterion, the dependence of R_θ and X_θ on θ must be determined experimentally. Although this criterion violates the normality principle, it provides a reasonably good fit with experimental data.

Zhou [14] proposed yet another nonquadratic anisotropic criterion.

$$\bar{\sigma}^{m} = [(3/2)/(F+G+H)]\left\{\left[\left(F(\sigma_y^2 + 3\tau_{xy}^2)^{m/2} + G\left(\sigma_x^2 + 3\,\tau_{xy}^2\right)^{m/2}\right.\right.\right.$$
$$\left.\left. + H\left[(\sigma_x - \sigma_y)^2 + 4\tau_{xy}^2\right]^{m/2} + 2N\tau_{xy}^2\right]^{m/2}\right\}. \qquad 7.43$$

This reduces to Hill's 1948 criterion for m = 2 and to equation 7.20 for $\tau_{xy} = 0$. Good agreement was found for m = 8. The yield stress in a 45 direction is

$$(X_{45}/X_0) = 2P(R+1)/[(R+P)(Q+1)]. \qquad 7.44$$

Still another non-quadratic criterion has been suggested [15]

$$c|\alpha_1\sigma_1 + \alpha_2\sigma_2|^m + h|\sigma_1 - \sigma_2|^m + 2n|\tau_{12}| = \sigma_0, \qquad 7.45$$

where σ_0 is the tensile yield strength, so there are five independent constants.

NOTE OF INTEREST

Rodney Hill was born June 11, 1921. He was a Professor of Mechanics of Solids at the University of Cambridge. Published in 1950, his book *The Mathematical Theory of Plasticity* forms the foundation of plasticity theory. He is among the foremost contributors to the foundations of solid mechanics over the second half of the twentieth century. His work is central to founding the mathematical theory of plasticity. He is also recognized world wide for his work's spare and concise style of presentation and for its exemplary standards of scholarship. In 1993, he won the Royal Medal for his contribution to the theoretical mechanics of soils and the plasticity of solids. He was elected a Fellow of the Royal Society in 1961. He died on February 2, 2011.

REFERENCES

1. R. Hill, *Proc. Roy. Soc.* v. 193A (1948).
2. R. Hill, *The Mathematical Theory of Plasticity*, Clarendon Press (1950).
3. W. F. Hosford, *The Mechanics of Crystals and Textured Polycrystals*, Oxford University Press (1993).
4. W. F, Hosford, "On Yield Loci of Anisotropic Cubic Metals," *7th North American Metalworking Res. Conf.* SME (1979).
5. R. Logan and W. F. Hosford, *Int. J. Mech. Sci.* v. 22 (1980).
6. R. Hill, *Math. Proc. Camb. Soc.* v. 75 (1979).
7. R. Pearce, *Sheet Metal Forming*, Adam Hilger (1991).
8. Y. Zhu, B. Dodd, R. M. Caddell, and W. F. Hosford, *Int. J. Mech. Sci.*, v. 29 (1987).
9. F. Barlat and J. Lian, *Int. J. Plasticity* v.5 (1989).
10. D. J. Lege, F. Barlat, and J. C. Brem, *Int. J. Mech Sci.* v. 31 (1989).
11. F. Barlat, D. J. Lege, and J. C. Brem, *Int. J. Plasticity*, v. 7 (1991).
12. M. Gotoh, *Int. J. Mech Sci.* v. 19 (1977).
13. W. F. Hosford, *Int. J. Mech Sci.* v. 27 (1985).
14. Zhou Wexian, *Int. J. Mech Sci.* v. 32 (1990).
15. F. Monttheillet, *Int. J. Mech Phys Solids.* v. 33 (1991).

8

SLIP AND DISLOCATIONS

Plastic deformation of crystalline materials usually occurs by *slip*, which is the sliding of planes of atoms slide over one another (Figure 8.1). The planes on which slip occurs are called *slip planes* and the directions of the shear are the *slip directions*. These are crystallographic planes and directions, and are characteristic of the crystal structure. The magnitude of the shear displacement by slip is an integral number of inter-atomic distances, so that the lattice is left unaltered by slip. If slip occurs on only a part of a plane, there remains a boundary between the slipped and unslipped portions of the plane, which is called a *dislocation*. It is the motion of these dislocations that cause slip.

Slip lines can be seen on the surface of deformed crystals. The fact that we can see these indicates that slip is inhomogeneous on an atomic scale. Displacements of thousands of atomic diameters must occur on discrete or closely spaced planes to create steps on the surface that are large enough to be visible. Furthermore, the planes of active slip are widely separated on an atomic scale. Yet the scale of the slip displacements and distances between slip lines are small compared to most grain sizes so slip usually can be considered as homogeneous on a macroscopic scale.

SLIP SYSTEMS

The slip planes and directions for several common crystals are summarized in Table 8.1. Almost without exception, the slip directions are

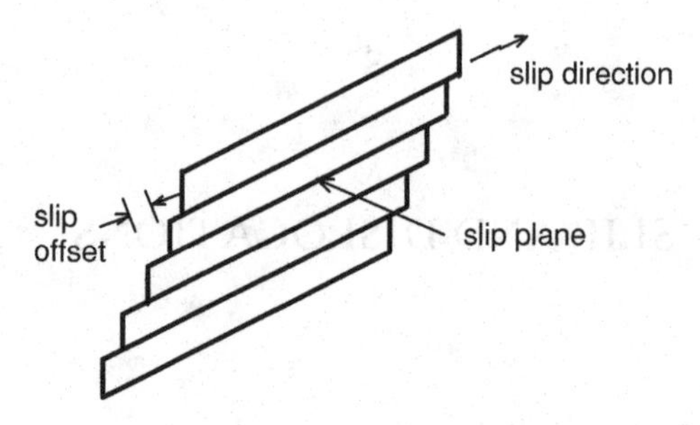

Figure 8.1. Slip by shear between parallel planes of atoms.

the crystallographic directions with the shortest the distance between like atoms or ions and the slip planes are usually densely packed planes.

SCHMID'S LAW

Schmid [1] realized that when a crystal is stressed, slip begins when the shear stress on a slip system reaches a critical value, τ_c, often called the *critical resolved shear stress*. In most crystals, slip occurs with equal

Table 8.1. *Slip directions and planes*

Structure	Slip direction	Slip planes
fcc	<110>	{111}
bcc	<111>	{110}, {112}, {123}, pencil glide*
hcp	$\langle 11\bar{2}0\rangle$	(0001), $\{1\bar{1}00\}$, $\{1\bar{1}01\}$
	$\langle 11\bar{2}3\rangle$†	$\{1\bar{1}0\bar{1}\}$
dia. cub.	<110>	{111}
NaCl	<110>	{110}
CsCl	<001>	{100}
Fluorite	<110>	{001}, {110}†, {111}†

* With pencil glide, slip is possible on all planes containing the slip direction.
† Slip has been reported under special loading conditions or high temperatures.

ease forward or backward, so the condition necessary for slip can be written as

$$\tau_{nd} = \pm\tau_c. \qquad 8.1$$

The subscripts n and d refer to the slip plane normal, n, and the direction of slip, d, respectively. This simple yield criterion for crystallographic slip is called Schmid's law. In a uniaxial tension test along the x-direction, the shear stress can be found from the stress transformation,

$$\tau_{nd} = \ell_{nx}\ell_{dx}\sigma_x, \qquad 8.2$$

where ℓ_{nx} and ℓ_{dx} are the cosines of the angles between the x and n directions and the x and d directions. Schmid's law is usually written as

$$\tau_c = \pm\sigma_x \cos\lambda \cos\phi, \qquad 8.3$$

where λ is the angle between the slip direction and the tensile axis, and ϕ is the angle between the tensile axis and the slip-plane normal (Figure 8.2), so $\ell_{dx} = \cos\lambda$ and $\ell_{nx} = \cos\phi$.

Equation 8.3 can be shortened to

$$\sigma_x = \pm\tau_c/m_x, \qquad 8.4$$

where $m_x = \cos\lambda \cos\phi$ is called the *Schmid factor*. The condition for yielding under a general stress state is

$$\pm\tau_c = \ell_{nx}\ell_{dx}\sigma_{xx} + \ell_{ny}\ell_{dy}\sigma_{yy} + \cdots\cdots\cdots + (\ell_{nx}\ell_{dy} + \ell_{ny}\ell_{dx})\sigma_{xy}. \qquad 8.5$$

Body-centered cubic crystals are exceptions. The stress required to cause slip in one direction may be different from that for the opposite direction. For the [111] direction, the stress to cause slip is greater in the $[2\bar{1}\bar{1}]$ than in the $[1\bar{1}0]$ direction as shown in Figure 8.3.

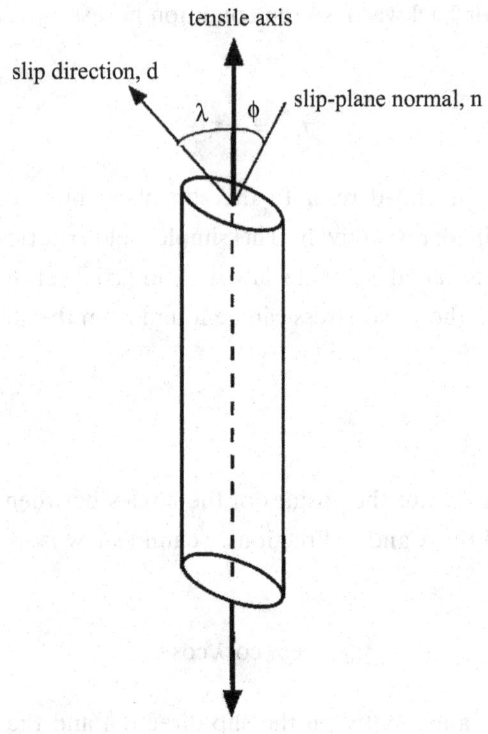

Figure 8.2. Slip elements in uniaxial tension. From W. F. Hosford, *Mechanical Behavior of Materials*, 2nd ed., Cambridge University Press (2010).

STRAINS PRODUCED BY SLIP

The incremental strain transformation equations may be used to find the shape change that results from slip when the strains are small (that is, when the lattice rotations are negligible). For infinitesimal strains,

$$d\varepsilon_{xx} = \ell_{xn}^2 d\varepsilon_{nn} + \ell_{xd}^2 d\varepsilon_{dd} + \cdots\cdots\cdot \ell_{xn}\ell_{xd} d\gamma_{nd}. \qquad 8.6.$$

For slip on a single nd slip system, the only strain term on the right-hand side of equation 8.6 is $d\gamma_{nd}$

$$d\varepsilon_{xx} = \ell_{xn}\ell_{xd} d\gamma_{nd}. \qquad 8.7$$

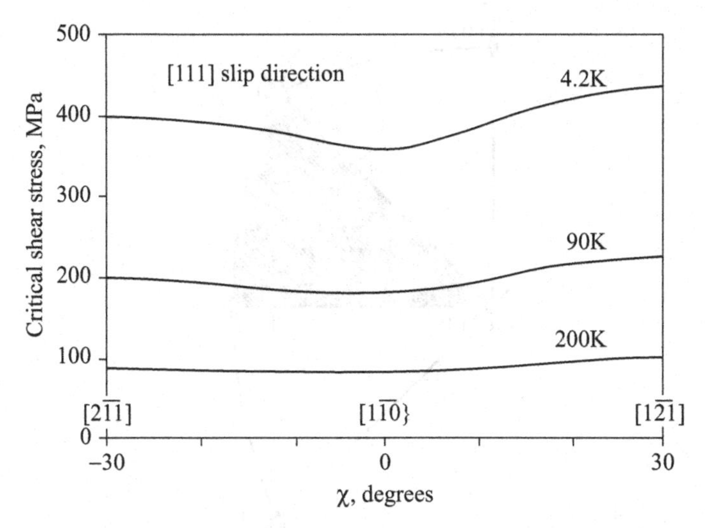

Figure 8.3 The critical stress for slip in the [111] direction of iron depends on the angle *X* between the slip plane normal and the [1$\bar{1}$0] direction. [1$\bar{1}$0]. From William F. Hosford, "Ferrite, Deformation and Fracture of," *Encyclopedia of Materials Science and Technology*, Elsevier Science Ltd, 2001.

In Schmid's notation, this is

$$d\varepsilon_{xx} = \cos\lambda\cos\phi d\gamma = m d\gamma, \quad 8.8$$

where $d\gamma$ is the shear strain on the slip system. The other strain components referred to the x, y, and z axes are similarly:

$$\begin{aligned}
d\varepsilon_{yy} &= \ell_{yn}\ell_{yd}d\gamma \\
d\varepsilon_{zz} &= \ell_{zn}\ell_{zd}d\gamma \\
d\gamma_{yz} &= (\ell_{yn}\ell_{zd} + \ell_{yd}\ell_{zn})d\gamma \\
d\gamma_{zx} &= (\ell_{zn}\ell_{xd} + \ell_{zd}\ell_{xn})d\gamma \\
d\gamma_{xy} &= (\ell_{xn}\ell_{yd} + \ell_{xd}\ell_{yn})d\gamma
\end{aligned} \quad 8.9$$

TENSILE DEFORMATION OF FCC CRYSTALS

It is customary to represent the tensile axis in the basic stereographic triangle with [100], [110], and [111] corners, as shown in Figure 8.4. For all orientations of fcc crystals within this triangle, the Schmid factor

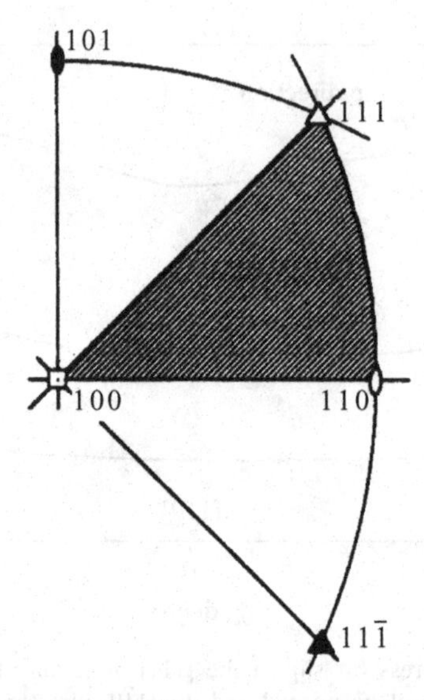

Figure 8.4. Basic orientation triangle. If the tensile axis of an fcc crystal lies in this triangle, the most heavily stressed slip system is [101](11$\bar{1}$). From W. F. Hosford, *Mechanical Behavior of Materials*, 2nd ed., Cambridge University Press (2010).

for slip in the [101] direction on the (11$\bar{1}$) plane is higher than for any other slip system. Slip on this *primary* slip system should be expected. If the tensile axis is represented as lying in any other stereographic triangle, the slip elements may be found by examining the remote corners of the three adjacent triangles. The <111> direction in one of the adjacent triangles is the normal to slip-plane and the <110> direction in another adjacent triangle is the slip direction.

Figures 8.5 and 8.6 show the orientation dependence of the Schmid factor within the basic triangle. The highest value, $m = 0.5$, occurs where the tensile axis lies on the great circle between the slip direction and the slip-plane normal with $\lambda = 45°$ and $\phi = 45°$.

If the tensile axis lies on a boundary of the basic triangle, two slip systems are equally favored. They are the ones most favored in the two triangles that form the boundary. At the corners there are four, six, or eight equally favored slip systems.

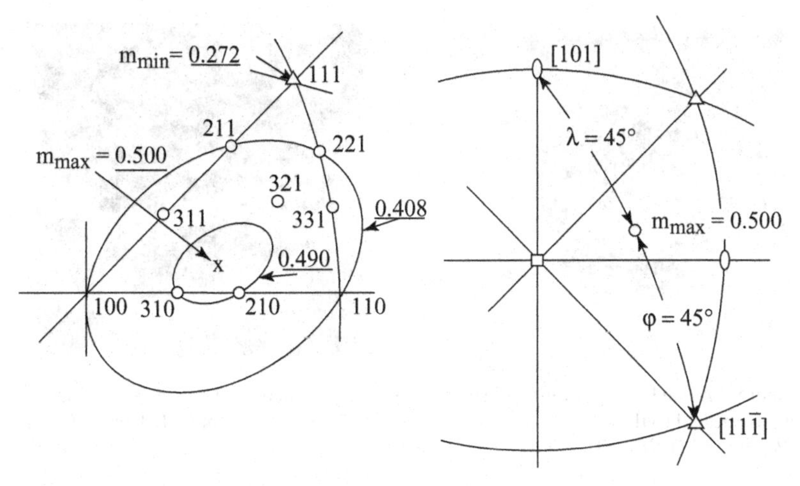

Figure 8.5. Orientation dependence of the Schmid factor for fcc crystals. From W. F. Hosford, *Mechanical Behavior of Materials*, 2nd ed., Cambridge University Press (2010).

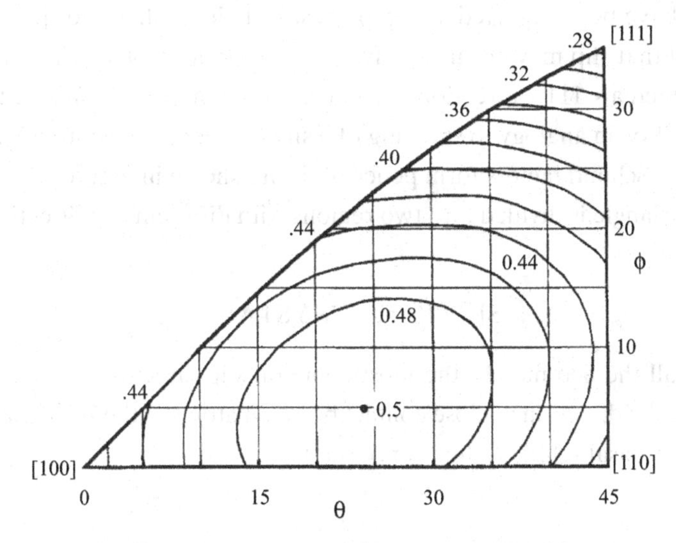

Figure 8.6. Full representation of contours of constant Schmid factor for fcc crystals. From W. F. Hosford, *Mechanical Behavior of Materials*, 2nd ed., Cambridge University Press (2010).

Figure 8.7. Taylor's illustration of shearing of a stack of pencils. The direction of shearing is parallel to the pencils but can take place on any plane. From G. I. Taylor, *J. Inst. Metals* v. 62 (1928).

SLIP IN bcc CRYSTALS

In bcc crystals, the slip direction is <111>. The {110}, {123}, and {112} have been reported as slip planes. In 1926, G. I. Taylor [2] suggested that slip may occur on any plane, crystallographic or not, that contained a <111> direction. He coined the term *pencil glide* for this possibility, in analogy to shearing of a stack of pencils (Figure 8.7).

The Schmid factors form pencil glide are shown in Figure 8.8. The basic triangle is divided into two regions with different slip directions.

SLIP IN hcp CRYSTALS

With all the hcp metals, the most common slip direction is $<11\bar{2}0>$. This is the direction of close contact between atoms in the basal plane, but non-basal $<11\bar{2}3>$ slip is important.

THEORETICAL STRENGTH OF CRYSTALS

It was well known in the late nineteenth century that crystals deformed by slip. However, it wasn't until the early twentieth century it was

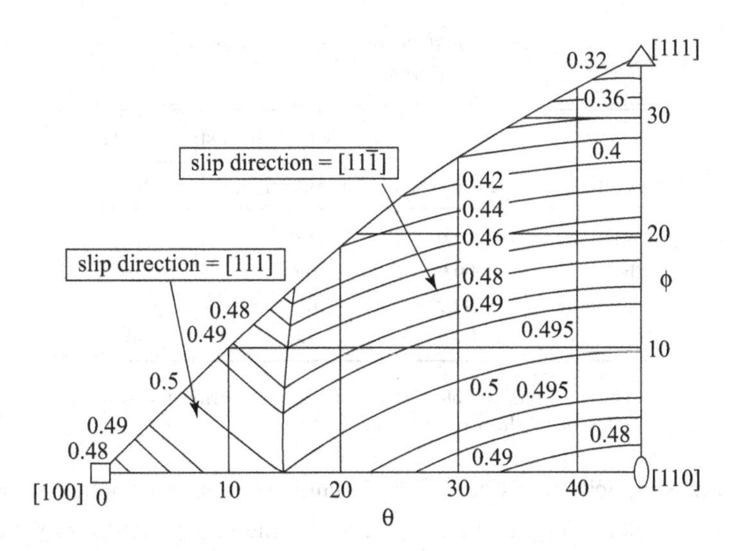

Figure 8.8. Plot of Schmid factors for <111>-pencil glide of bcc crystals. The basic triangle is divided into two regions with different slip directions. The Schmid factor is 0.417 at [100 and [110] and it is 0.314 at [111]. From W. F. Hosford, *The Mechanics of Crystals and Textured Polycrystals*, Oxford University Press (1993).

realized that the stresses required to cause slip were much lower than theory predicted.

Once it was established that crystals deformed by slip on specific crystallographic systems, physicists tried to calculate the strength of crystals. However, calculated strengths were far in excess of the experimental measurements. The predicted strengths were two orders of magnitude too high as indicated in Table 8.2.

The basis for the theoretical calculations is illustrated in Figure 8.9. Each plane of atoms nestles in pockets formed by the plane below (a). As a shear stress is applied to a crystal, the atoms above the plane of shear must rise out of the stable pockets in the plane below, and slide over it until they reach the unstable position shown in (b). From this point, they will spontaneously continue to shear to the right until they reach a new stable position (c).

Table 8.2. *Critical shear stress for slip in several materials*

		Critical shear stress (MPa)	
Metal	Purity %	Experiment*	Theory
Copper	>99.9	1.0	414
Silver	99.99	0.6	285
Cadmium	99.996	0.58	207
Iron	99.9	7.0	740

* There is considerable scatter caused by experimental variables, particularly purity.

For simplicity, consider a simple cubic crystal. An applied shear stress, τ, will displace one plane relative to the next plane as shown in Figure 8.10.

When the shear displacement, x, is 0, d, $2d$, or nd (i.e., $\gamma = 0, 1, 2, \ldots n$), the lattice is restored, so t should be zero. The shear stress, t, is also zero when the displacement is $x = (1/2)d$, $(3/2)d$, etc. ($\gamma = 1/2, 3/2, \ldots.$). A sinusoidal variation of τ with γ as shown in Figure 8.11 seems reasonable, so

$$\tau = \tau_{\max}\sin(2\pi\gamma). \qquad 8.10$$

Here $\tau_{\max}$ is the theoretical shear stress required for slip. If the stress is less than $\tau_{\max}$, the shear strain is elastic, and will disappear when the stress is released. For very low values of γ (Figure 8.12), Hooke's law

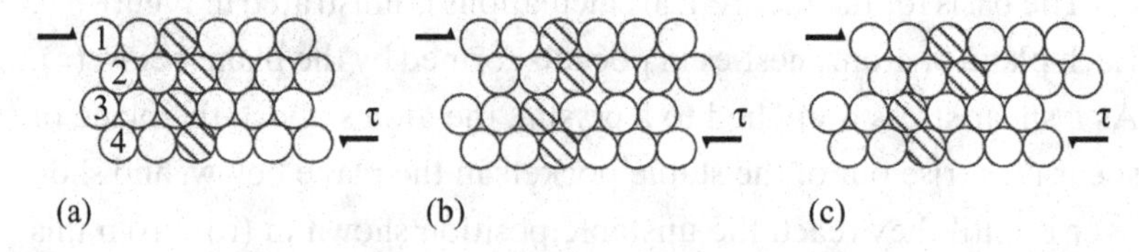

Figure 8.9. Model of slip occurring by sliding of planes 1 and 2 over planes 3 and 4. At the unstable condition (b) the planes are at attracted equally to the stable configurations in (a) and (c). From W. F. Hosford, *The Mechanics of Crystals and Textured Polycrystals*, Oxford University Press (1993).

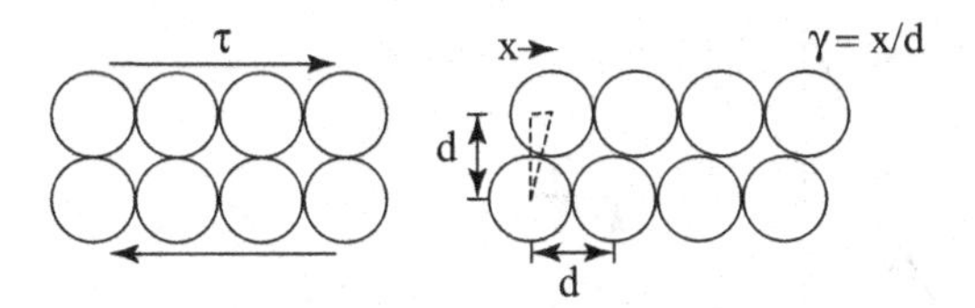

Figure 8.10. Model used to calculate the theoretical shear stress for slip. From W. F. Hosford, *The Mechanics of Crystals and Textured Polycrystals*, Oxford University Press (1993).

should apply,

$$\tau = G\gamma. \quad 8.11$$

This can be expressed as

$$(d\tau/d\gamma)_{\gamma\to 0} = G. \quad 8.12$$

Differentiating equation 8.10,

$$(d\tau/d\gamma)_{\gamma\to 0} = 2\pi\tau_{max}\cos(2\pi\gamma)_{\gamma\to 0} = 2\pi\tau_{max}.$$

$$\tau_{max} = G/2\pi. \quad 8.13$$

A somewhat more sophisticated analysis for real crystal structures predicts something close to

$$\tau_{max} = G/10. \quad 8.14$$

In Table 7.2, the theoretical strength values predicted by equation 8.5 are orders of magnitude higher than experimental measurements. The poor agreement between experimental and theoretical strengths indicated that there was something wrong with the theory. The

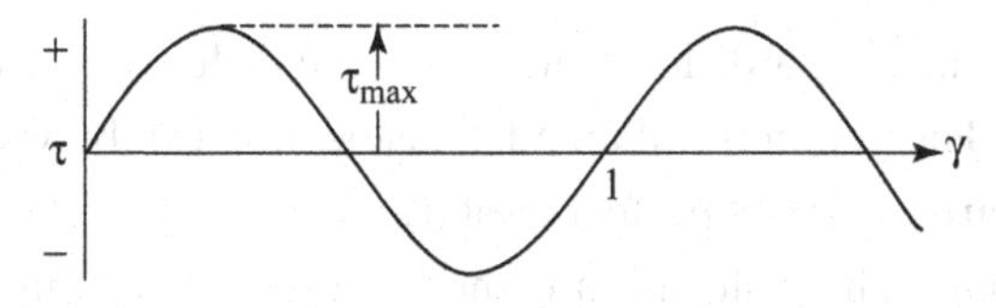

Figure 8.11. Theoretical variation of shear stress with displacement. From W. F. Hosford, *The Mechanics of Crystals and Textured Polycrystals*, Oxford University Press (1993).

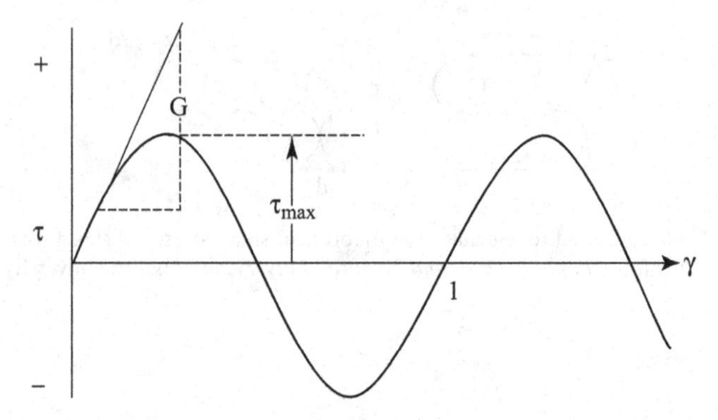

Figure 8.12. The shear modulus, G, is the initial slope of the theoretical τ vs. γ curve. From W. F. Hosford, *The Mechanics of Crystals and Textured Polycrystals*, Oxford University Press (1993).

problem with the theoretical calculations is that it was assumed that slip occurs by one *entire plane* of atoms sliding over another at the same time. Taylor [3], Orowan [4], and Polanyi [5] realized that it is not necessary for an entire plane to slip at the same time. They postulated that crystals have pre-existing defects that are boundaries between regions that are already displaced relative to one another by a unit of slip. These boundaries are called *dislocations*. It is the movement of dislocations that causes slip. The critical stress for slip is the stress required to move a dislocation. At any instant, shearing occurs at the dislocation rather than over the entire slip plane. It was another two decade before dislocations were directly observed.

THE NATURE OF DISLOCATIONS

One special form of a dislocation is an *edge* dislocation sketched in Figure 8.13. The geometry of an edge dislocation can be visualized as having cut part way into a perfect crystal and then inserted an extra half plane of atoms. The dislocation is the bottom edge of this extra half plane. The *screw dislocation* (Figure 8.14) is another special form. It is like a spiral-ramp parking structure. One circuit around the dislocation

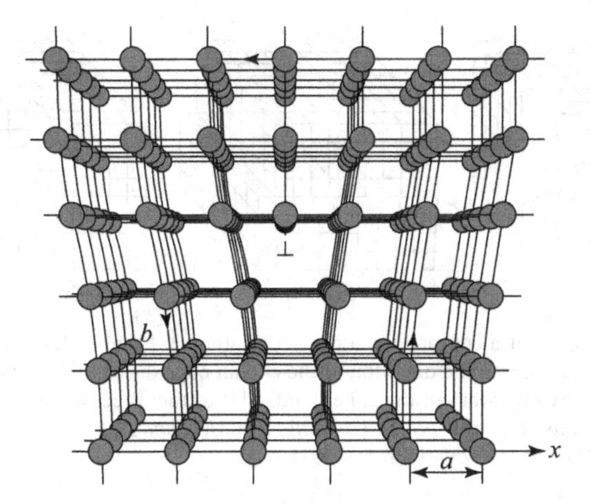

Figure 8.13. An edge dislocation is the edge of an extra half plane of atoms. From A. G. Guy, *Elements of Physical Metallurgy*, Addison-Wesley, (1951).

leads one plane up or down. Planes are connected in a manner similar to the levels of a spiral-parking ramp.

An alternate way of visualizing dislocations is illustrated in Figure 8.15. If a crystal is cut, an edge dislocation is created by shearing the top half of the crystal by one atomic distance perpendicular to the end of the cut (Figure 8.14b). This produces an extra half plane of atoms,

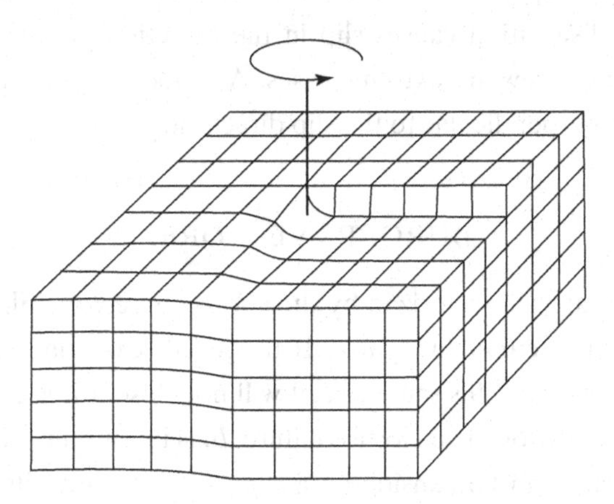

Figure 8.14. A screw dislocation. Traveling on the lattice around the screw dislocation is like following the thread of a screw. From W. F. Hosford, *The Mechanics of Crystals and Textured Polycrystals*, Oxford University Press (1993).

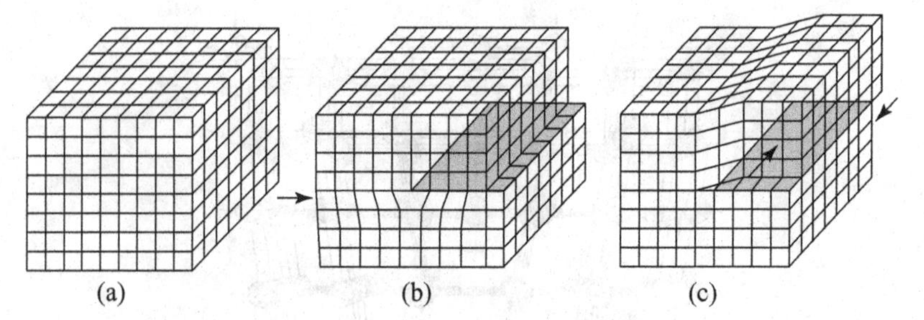

Figure 8.15. Consider a cut made in a perfect crystal (a). If one half is sheared by one atom distance parallel to the direction of the cut, an edge dislocation results (b). If one half is sheared by one atom distance perpendicular to the direction of the cut, a screw dislocation results (c). From W. F. Hosford, *The Mechanics of Crystals and Textured Polycrystals*, Oxford University Press (1993).

the edge of which is the center of the dislocation. The other extreme form of a dislocation is the *screw dislocation.* A screw dislocation is generated by cutting into a perfect crystal and then shearing half of it by one atomic distance in a direction parallel to the end of the cut (Figure 8.14c). The end of the cut is the dislocation.

In both cases, the dislocation is a boundary between regions that have and have not slipped. When an edge dislocation moves, the direction of slip is perpendicular to the dislocation. In contrast, movement of a screw dislocation causes slip in the direction parallel to itself. The edge and screw are extreme cases. A dislocation may be neither parallel nor perpendicular to the slip direction.

BURGERS VECTORS

Dislocations are characterized by their *Burgers vectors* [6]. Consider an atom-to-atom circuit in Figure 8.16 that would close on itself if made in a perfect crystal. This same circuit will not close if it is constructed around a dislocation. The closure failure, b, is its Burgers vector. The Burgers vector can be considered a *slip vector* because its direction is the slip direction and its magnitude is the magnitude of the slip displacement caused by its movement of the dislocation.

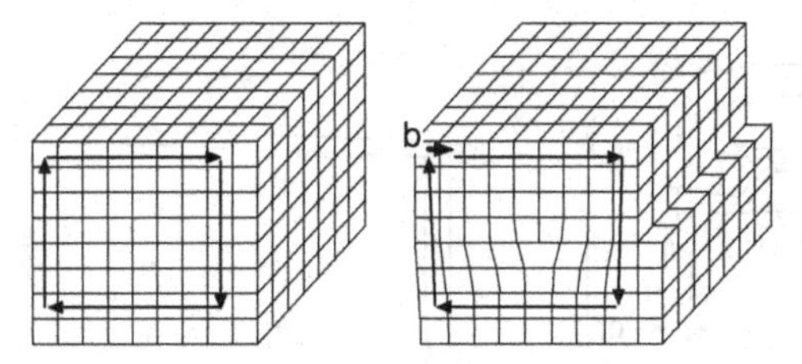

Figure 8.16. The Burgers vector of a dislocation can be determined by drawing a clockwise circuit that would close if it were drawn in a perfect crystal. If the circuit is drawn around a dislocation, the closure failure is the Burgers vector. From W. F. Hosford, *The Mechanics of Crystals and Textured Polycrystals*, Oxford University Press (1993).

A dislocation may wander through a crystal with its orientation changing from place to place, but everywhere its Burgers vector is the same (see Figure 8.17). If the dislocation branches into two dislocations, the sum of the Burgers vectors of the branches equals its Burgers vector.

In a simple notation system for describing the magnitude and direction of the Burgers vector of a dislocation, the direction is indicated by direction indices and the magnitude by a scalar preceding the direction. For example, $b = (a/3)[2\bar{1}1]$ in a cubic crystal means that the Burgers has components of $2a/3$, $-a/3$, and $a/3$ along the [100], [010] and [001] directions respectively, where a is the lattice parameter. Its magnitude is $|b| = [(2a/3)^2 + (-a/3)^2 + (a/3)^2]^{1/2} = a\sqrt{6}/3)$.

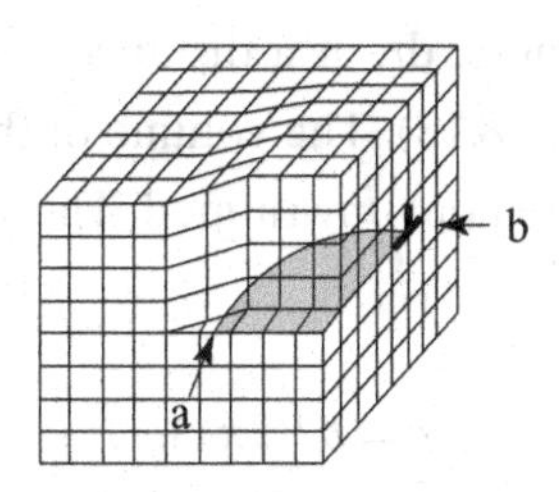

Figure 8.17. A dislocation may wander through a crystal, but everywhere it has the same Burgers vector. Where it is parallel to its Burgers vector, it is a screw (a). Where it is perpendicular to its Burgers vector, it is an edge (b). Elsewhere it has mixed character. W. F. Hosford, *The Mechanics of Crystals and Textured Polycrystals*, Oxford University Press (1993).

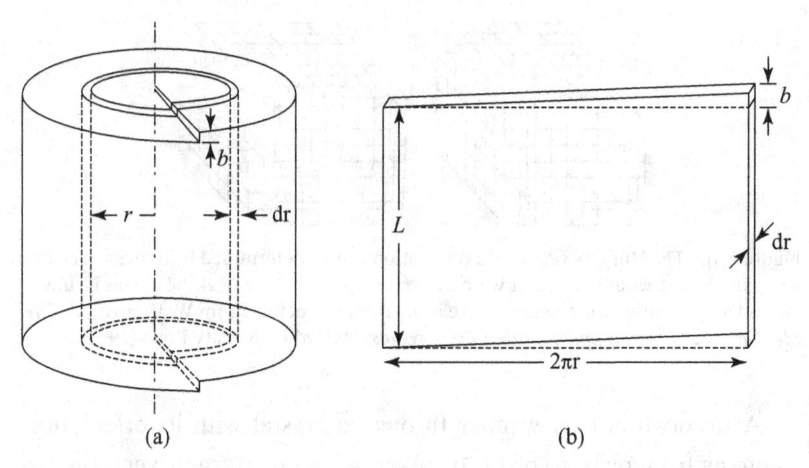

Figure 8.18 a. Screw dislocation in a cylindrical crystal; b. Flattened element. From W. F. Hosford, *The Mechanics of Crystals and Textured Polycrystals*, Oxford University Press (1993).

A dislocation in an fcc crystal corresponding to a full slip displacement that restores the lattice would have a Burgers vector $(a/2)<110>$. In this case, the magnitude is $a\sqrt{2}/2$.

ENERGY OF A SCREW DISLOCATION

The energy associated with a dislocation is the elastic energy stored in the lattice surrounding the dislocation. The distortion is severe near the dislocation but decreases with distance from it. Consider an element of length, L, and thickness, dr, at a distance, r, from the center of a screw dislocation (Figure 8.18). The volume of this element is $2\pi rL\mathrm{d}r$. Figure 8.18 shows this element unwrapped. The shear strain associated with this element is

$$\gamma = b/(2\pi r). \qquad 8.15$$

The energy/volume associated with an elastic distortion is

$$U_{\mathrm{v}} = (1/2)\tau\,\gamma, \qquad 8.16$$

where τ is the shear stress necessary to cause the shear strain, γ. According to Hooke's Law,

$$\tau = G\gamma, \qquad 8.17$$

where G is the shear modulus. Combining equations 8.15, 8.16, and 8.17, the energy per volume is

$$U_v = G\gamma^2/2 = (1/2)Gb^2/(2\pi r)^2. \qquad 8.18$$

The elastic energy, dU, associated with the element is its energy/volume times its volume,

$$\mathrm{d}U = [(1/2)Gb^2/(2\pi r)^2](2\pi rLdr) = \mathrm{G}b^2L/(2\pi r)\mathrm{d}r. \qquad 8.19$$

The total energy of the dislocation per length, L, is obtained by integrating equation 8.19.

$$U/L = [Gb^2/(4\pi)]\int_{r_0}^{r_1} dr/r = [Gb^2/(4\pi)]\ln(r_1/r_0) \qquad 8.20$$

If the lower limit, r_o, of integration is taken as zero, the value of U in equation 7.20 would be infinite. The problem is that equation 8.20 predicts an infinite strain at $r = 0$, which corresponds to an infinite energy per volume at the core of the dislocation, ($r_o = 0$). This is clearly unreasonable. The energy per volume cannot be higher than the heat of vaporization. This discrepancy arises because Hooke's law breaks down at the very high strains that exist near the core of the dislocation. The stress, τ, is proportional to G only for small strains. A lower limit, r_o, can be chosen so that the neglected energy of the core in the region, $0 < r < r_o$, is equal to the overestimation of the integral for $r > r_o$ (see Figure 8.19). A value of $r_o = b/4$ has been suggested as reasonable.

The upper limit of integration, r_1, can't be any larger than the radius of the crystal. A reasonable value for r_1 is half of the distance between dislocations, the stress field is neutralized by stress fields of other dislocations. The value of r_1 is often approximated by 10^5b, which corresponds to a dislocation density of 10^{10} dislocations per m^2.

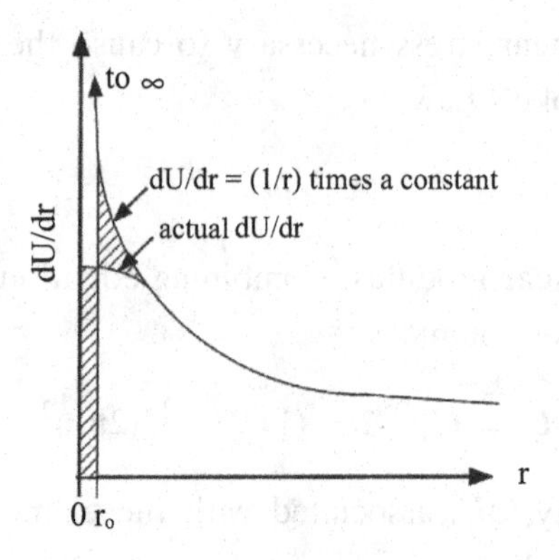

Figure 8.19. Energy density near core of a screw dislocation. From W. F. Hosford, *The Mechanics of Crystals and Textured Polycrystals*, Oxford University Press (1993).

This is convenient because $\ln(10^5 b/0.25b) = \ln(4 \times 10^5) = 12.9 \approx 4\pi$. With this approximation, the energy per length, U_L, from equation 8.20 simplifies to

$$U_L \approx Gb^2. \tag{8.21}$$

The process of choosing the limits of integration so that the expression for the energy per length simplifies to equation 8.21 may seem arbitrary. Yet the results, although approximate, are reasonable. The derivation of the energy of an edge dislocation is more complicated because the stress field around an edge dislocation is more complex. For edge dislocations,

$$U_L \approx Gb^2/(1-\upsilon) \tag{8.22}$$

where υ is Poisson's ratio. Thus, the energy of an edge dislocation is greater than that of a screw by a factor of $1/(1-\upsilon) \approx 1.5$.

There are two important features of equations 8.21 and 8.22. One is that the energy of a dislocation is proportional to its length. Energy per length is equivalent to line tension, or a contractile force. The units of energy/length are J/m, which is the same as the units of force, N. The second important feature is that the energy of a dislocation is proportional to b^2. This controls the energetics of reactions between parallel dislocations.

REACTIONS BETWEEN PARALLEL DISLOCATIONS

Two parallel dislocations may combine and form a third dislocation. If they do, the Burgers vector of the third dislocation will be the vector sum of the Burgers vectors of the two reacting dislocations. That is, if $b_1 + b_2 \rightarrow b_3$, then $b_1 + b_2 = b_3$. Such a reaction is energetically favorable if it lowers the energy of the system. Frank's rule states that since the energy of a dislocation is proportional to b^2, the reaction is favorable if $b_1^2 > b_2^2 + b_3^2$. Similarly, a dislocation, b_1, may spontaneously dissociate into two parallel dislocations, b_2 and b_3, if $b_1^2 + b_2^2 > b_3^2$. It is energetically favorable for dislocations with large Burgers vectors to react with one another to form dislocations with smaller Burgers vectors. As a consequence, dislocations tend to have small Burgers vectors.

STRESS FIELDS AROUND DISLOCATIONS

Atoms near a dislocation are displaced from their normal lattice positions. These displacements are equivalent to the displacements caused by elastic strains arising from external stresses. In fact, the lattice displacements or strains can be completely described by these stresses and Hooke's laws can be used to find them.

It should be noted that the equations given below are based on the assumption of isotropic elasticity. For a screw dislocation with a Burgers vector, b, parallel to the z axis,

$$\tau_{z\theta} = -Gb/(2\pi r) \quad \text{and} \quad \tau_{r\theta} = \tau_{rz} = \sigma_r = \sigma_\theta = \sigma_z = 0, \qquad 8.23$$

where G is the shear modulus and r is the radial distance from the dislocation. The minus sign indicates that the repulsive force is inversely proportional to the distance between the dislocations.

Equation 8.23 indicates that a screw dislocation creates no hydrostatic tension or compression, σ_H, because $\sigma_H = (\sigma_r + \sigma_\theta + \sigma_z)/3$. Therefore, there should be no dilatation (volume strain) associated with a screw dislocation. (However real crystals are elastically anisotropic so there may be small dilatations associated with screw dislocations.)

For an edge dislocation, which lies parallel to z and has its Burgers vector parallel to x:

$$\tau_{xy} = Dx(x^2 - y^2)/(x^2 + y^2)^2, \quad 8.24a$$

$$\sigma_{xy} = Dy(3x^2 + y^2)/(x^2 + y^2)^2, \quad 8.24b$$

$$\sigma_y = Dy(x^2 - y^2)/(x^2 + y^2)^2 \quad 8.24c$$

$$\sigma_z = \upsilon(\sigma_x + \sigma_y) = -2D\upsilon y/(x^2 + y^2), \tau_{yz} = \tau_{zx} = 0, \quad 8.24d$$

where $D = Gb/[2\pi(1 - \upsilon)]$.

One of the important features of these equations is that there is a hydrostatic stress, $\sigma_H = (\sigma_x + \sigma_y + \sigma_z)/3$, around an edge dislocation. Combining equations 8.24b, c, and d, $\sigma_H = -(2/3)Dy(1 + \upsilon)/(x^2 + y^2)$ or

$$\sigma_H = -A(y)/(x^2 + y^2), \quad 8.25$$

where $A = Gb(1 + \upsilon)/[3\pi(1 - \upsilon)]$. Figure 8.20 shows how the hydrostatic stress varies near an edge dislocation. There is hydrostatic compression (negative σ_H) above the edge dislocation (positive y) and hydrostatic tension below it.

In substitutional solutions, solute atoms that are larger than the solvent atoms are attracted to regions just below edge dislocations where their larger size helps relieve the hydrostatic tension. Similarly, substitutional solute atoms that are smaller than the solvent atoms are attracted to regions just above edge dislocations. In either case, edge dislocation will attract solute atoms (Figure 8.19). In interstitial solid solutions, the solute atoms are attracted to the region just below the

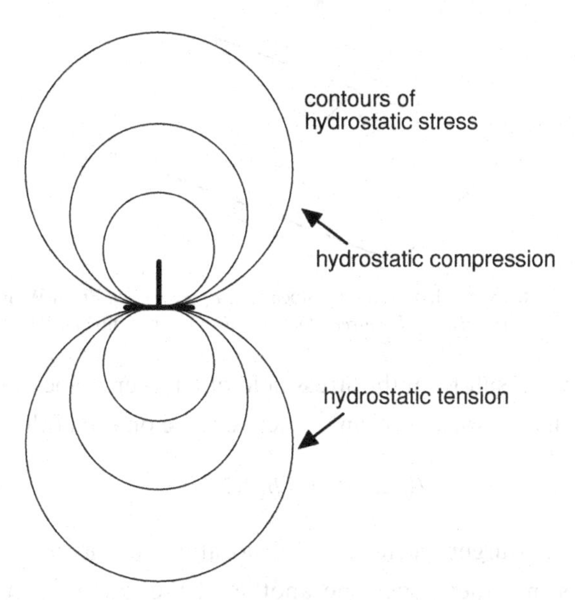

Figure 8.20. Contours of hydrostatic stress around an edge dislocation. Note that the level of hydrostatic stress increases near the dislocation. From W. F. Hosford, *The Mechanics of Crystals and Textured Polycrystals*, Oxford University Press (1993).

edge dislocation where they help relieve the tension. It is this attraction of edge dislocations in iron for carbon and nitrogen that is responsible for the yield point effect and strain-aging phenomenon in low carbon steel.

FORCES ON DISLOCATIONS

Stresses in crystals cause forces on dislocations. Consider a dislocation of length, L, and Burgers vector, b, on a plane as shown in Figure 8.21. A shear stress, τ, acting on that plane will cause a force on the dislocation, per unit length, F_L, of

$$F_L = -\tau \cdot b. \tag{8.26}$$

A dot product is possible in equation 8.25 because once the plane of the stress is fixed, the stress can be treated as a vector (force). The

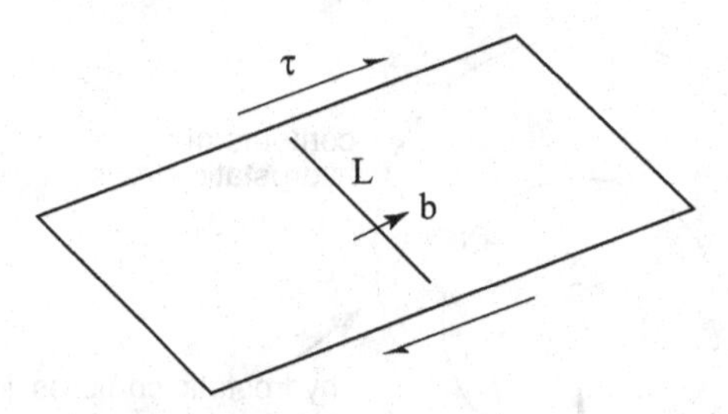

Figure 8.21. The force per length on a dislocation, $F_L = -\tau \cdot b$. From W. F. Hosford, *The Mechanics of Crystals and Textured Polycrystals*, Oxford University Press (1993).

stress, τ, may result from the stress field of another dislocation. Thus, two screw dislocations exert an attractive force on each other of

$$F_L = -Gb_1 \cdot b_2/(2\pi r), \qquad 8.27$$

where b is the Burgers vector of the dislocation of concern. The minus sign means that they repel one another if the dot product is positive. An equivalent statement is "two dislocations repel each other if Frank's rule predicts that their combination would result in an energy increase." If the angle between b_1 and b_2 is less than 90 degrees, $|b_1 + b_2| > |b_1| + |b_2|$.

The interaction of two parallel edge dislocations is somewhat more complex. The shear stress field for one edge dislocation that lies parallel to z with a Burgers vector parallel to x is given by equation 8.24a, $\tau_{xy} = Dx(x^2 - y^2)/(x^2 + y^2)^2$, where $D = Gb/[2\pi((1 - \upsilon)]$. The mutual force on that plane is

$$F_L = -\{b_1 \cdot b_2/[2\pi(1 - \upsilon)]\}x(x^2 - y^2)/(x^2 + y^2)^2. \qquad 8.28$$

For dislocations with like sign ($b_1 \cdot b_2 > 0$), there is mutual repulsion in the region x > y and attraction in the region $x < y$. This is equivalent to saying that there is mutual repulsion if Frank's rule predicts a reaction would cause an increase of energy and mutual attraction if it would cause a decrease of energy. Figure 8.22 shows the regions of attraction and repulsion.

The stress, τ_{xy}, is zero at $x = 0$, $x = y$ and $x = \infty$.

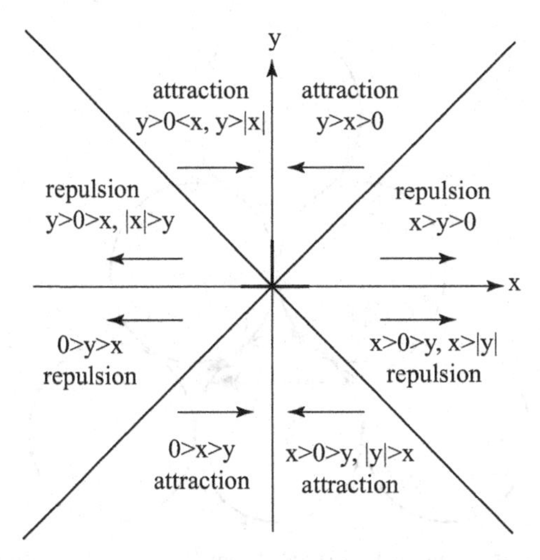

Figure 8.22. Stresses around an edge dislocation either attract or repel another parallel dislocation having the same Burgers vector, depending on how the two are positioned relative to one-another. From W. F. Hosford, *The Mechanics of Crystals and Textured Polycrystals*, Oxford University Press (1993).

Between $x = 0$ and $x = y$, τ_{xy} is negative, indicating that the stress field would cause attraction of edge dislocations of the same sign to each other. Therefore, edge dislocations of the same sign tend to line up one above the other to form low angle grain boundaries. For x greater than y, the stress, τ_{xy}, is positive indicating that the stress field would tend to repel another edge dislocation of the same sign.

PARTIAL DISLOCATIONS IN fcc CRYSTALS

In fcc crystals, slip occurs on {111} planes and in <110> directions. For the specific case of $(111)[\bar{1}10]$ slip, the Burgers vector which corresponds to displacements of one atom diameter is $(a/2)[\bar{1}10]$. A dislocation with this Burgers vector can dissociate into two partial dislocations,

$$(a/2)[\bar{1}10] \rightarrow (a/6)[\bar{2}11] + (a/6)[\bar{1}2\bar{1}]. \quad 8.29$$

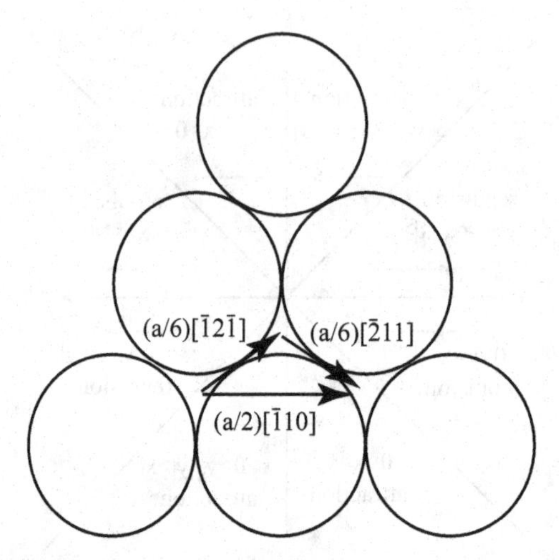

Figure 8.23. Slip systems in an fcc crystal. From W. F. Hosford, *The Mechanics of Crystals and Textured Polycrystals*, Oxford University Press (1993).

This reaction is vectorially correct because $b_1 = (-a/2, a/2, 0)$ does equal $b_2 = (-a/3, a/6, a/6) + b_3 = (-a/6, a/3, -a/6)$. Figure 8.23 is a geometrical representation of these two partials.

STACKING FAULTS

If a single $(a/6)<112>$ partial dislocation passes through an fcc crystal, it leaves behind a region in which the sequence of stacking of the close-packed {111} planes does not correspond to the normal fcc lattice. The correct stacking order is not restored until the second partial dislocation passes. The normal stacking order in fcc and hcp lattices is shown in Figure 8.24. In hcp crystals, the third close-packed plane, A, lies directly over the first, A. In fcc crystals, the third close-packed plan, C, is over neither the first (A) nor the second (B){111} plane. Figure 8.25 shows that a $(a/6)<112>$ partial dislocation changes the position of the third plane so that it is directly over the first and therefore produces a local region of hcp packing.

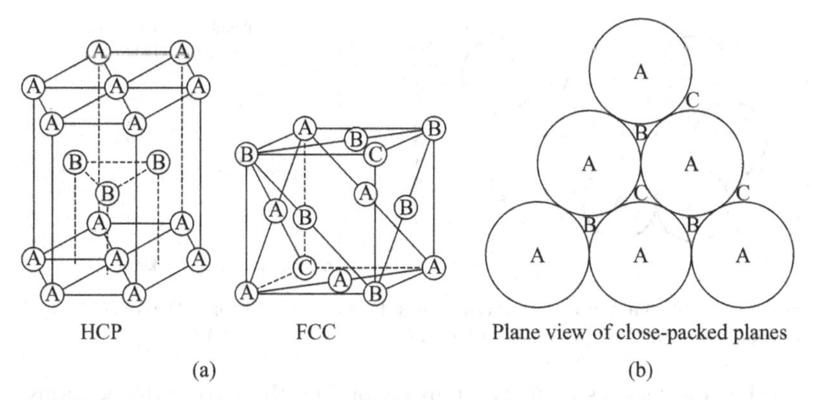

Figure 8.24 a. The normal stacking of {111} planes in an fcc crystal can be described as ABCABC. b. When a $(a/6)<112>$ partial dislocation passes through the crystal, the stacking sequence is changed to ABABC. From W. F. Hosford, *The Mechanics of Crystals and Textured Polycrystals*, Oxford University Press (1993).

In Figure 8.26, the stacking order near a stacking fault in an fcc crystal is compared with the stacking in fcc and hcp lattices and near a twin boundary in an fcc crystal.

The stacking sequence near a stacking fault in an fcc crystal is similar to the packing sequence in the hcp lattice. Since this is not the equilibrium structure of a fcc crystal, the stacking fault raises the energy and the increase of energy depends directly on the area of the fault. The stacking fault energy, γ_{SF}, is the energy per area of fault and

Figure 8.25. Stacking of close-paced planes in fcc and hcp crystals. When a $(a/6)<112>$ partial dislocation moves plane C so that it is directly over plane A, it creates a region where the packing sequence is hcp rather than fcc. From W. F. Hosford, *The Mechanics of Crystals and Textured Polycrystals*, Oxford University Press (1993).

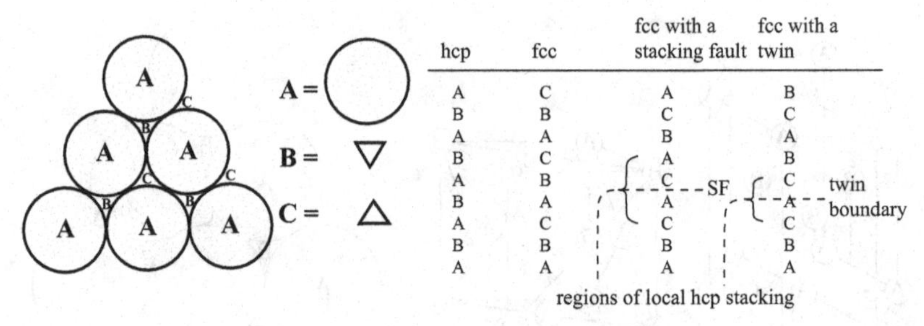

Figure 8.26. Stacking of close packed planes. From W. F. Hosford, *The Mechanics of Crystals and Textured Polycrystals*, Oxford University Press (1993).

can be regarded as a surface tension pulling the partial dislocations together. A stacking fault has twice as many incorrect second-nearest neighbors as a twin boundary. The similarity of the packing sequences at a twin boundary and at a stacking fault is clear in Figure 8.26. The frequency of annealing twins is much higher in fcc metals of low stacking fault energy (Ag, brass, Al-bronze, γ-stainless) than in those with higher stacking fault energy (Cu, Au). Annealing twins in the microstructures of aluminum alloys are rare. Table 8.3 lists the values of the stacking fault energy for a few fcc metals. The values of γ_{SF} for brass (Cu-Zn), aluminum bronze (Cu-Al), and austenitic stainless steel are still lower than the value for Ag.

Figure 8.27 illustrates a stacking fault and its two partial disocations.

FRANK-READ SOURCES

Once dislocation theory was proposed, two questions arose. First, why does the dislocation density increase with deformation and secondly, why are the offsets of the slip lines at the surface correspond to a large

Table 8.3. *Stacking fault energies of several fcc metals*

Ag	Al	Au	Cu	Ni	Pd	Pt	Rh	Ir
16	166	32	45	125	180	322	750	300 mJ/m^2.

From a listing in J.P. Hirth and J. Löther, *Theory of Dislocations*, Wiley 1982.

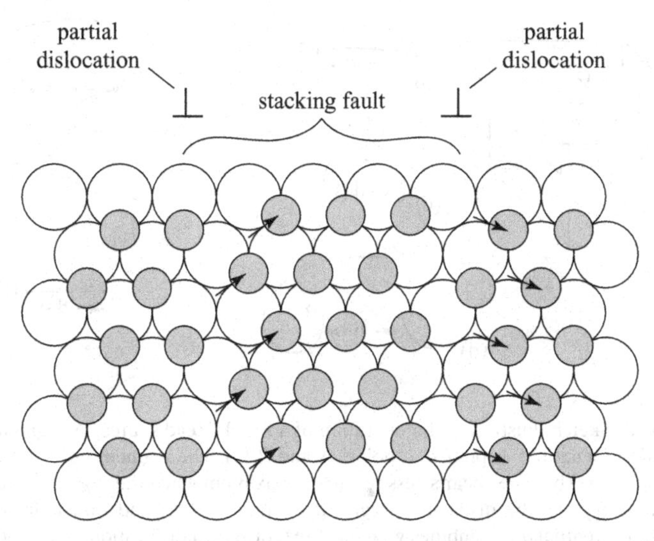

Figure 8.27. The equilibrium spacing between two partial dislocations corresponds to a balance of the mutual repulsion of the two partials and the surface energy of the stacking fault between them. From W. F. Hosford, *The Mechanics of Crystals and Textured Polycrystals*, Oxford University Press (1993).

number of atom diameters? These two questions can be answered in terms of the *Frank-Read source*, which generates dislocations. Suppose that there is a finite length of a dislocation, AB, in a slip plane (Figure 8.28). The dislocation leaves the plane at A and B, but the end points are pinned at A and B. A shear stress, τ, acting on the plane, will create a force that causes dislocation to bow. This bowing is resisted by the line tension of the dislocation. As the shear stress is increased, the dislocation will bow out until it spirals back on itself. The sections that touch annihilate each other, leaving a dislocation loop that can expand under the stress and a restored dislocation segment between the pinning points. The process can repeat itself producing many loops. Frank-Read sources explain how the number of dislocations increases with deformation. Their existence on some planes and not others, explains why slip occurs with large offsets (corresponding to the passing of many dislocations) on widely spaced planes.

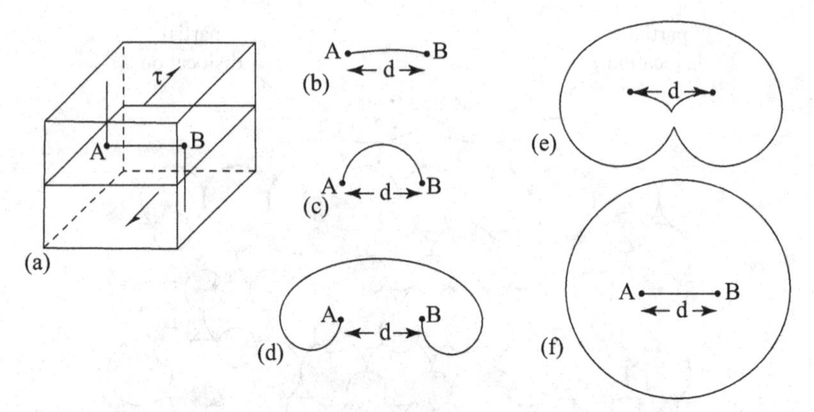

Figure 8.28. Sketch illustrating the operation of a Frank-Read source. A segment of a dislocation of length, d, is pinned at points A and B (a). The segment bows out under a shear stress, τ, (b). The shear stress reaches a maximum when the segment becomes a semicircle (c). As the dislocation segment continues to expand under decreasing stress (d), it eventually recombines with itself (e) forming a dislocation loop. The loop continues to expand and a new dislocation segment, AB, is formed. The process can repeat itself sending out many loops. From W. F. Hosford, *The Mechanics of Crystals and Textured Polycrystals*, Oxford University Press (1993).

The stress necessary to operate a Frank-Read source can be calculated by considering a balance of forces acting on the bowed segment of the dislocation (Figure 8.29). The applied shear stress, τ, causes a force, τbd, where d is the distance between the pinned ends. The line tension of the dislocation (energy per length), $U_L \approx Gb^2$ (equation 8.21), acts parallel to the dislocation line and tends to keep it from moving. Considering both ends, the vertical component of this force is $2Gb^2\sin\theta$. This force reaches a maximum, $2Gb^2$, when the dislocation

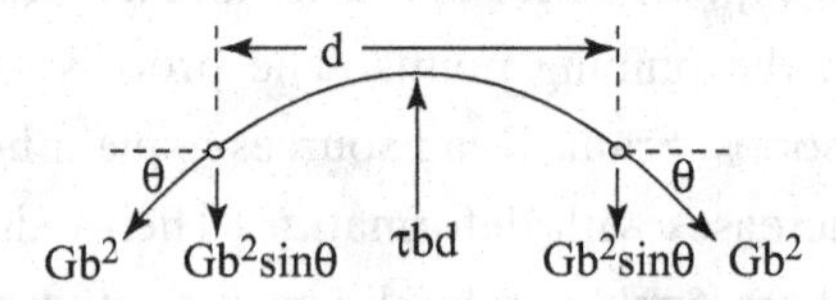

Figure 8.29. Force balance on a pinned dislocation that is bowed by a shear stress, τ. From W. F. Hosford, *The Mechanics of Crystals and Textured Polycrystals*, Oxford University Press (1993).

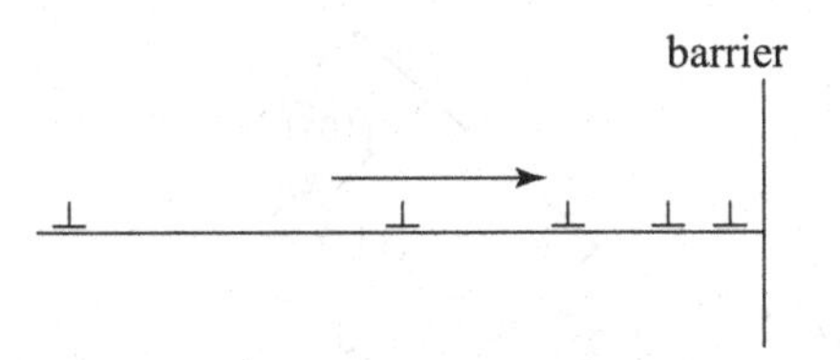

Figure 8.30. Pile-up of dislocations at an obstacle. From W. F. Hosford, *The Mechanics of Crystals and Textured Polycrystals*, Oxford University Press (1993).

is bowed into a semicircle ($\theta = 90°$). Equating these two forces and assuming the shear stress is parallel to b, $\tau bd = 2Gb^2$ or

$$\tau = 2Gb/d. \qquad 8.30$$

Thus, the stress necessary to operate a Frank-Read source is inversely proportional to the size of the source, *d*.

DISLOCATION PILE-UPS

When dislocations from a Frank-Read source come to an obstacle such as a grain boundary or hard particle, they tend to form a pile-up (Figure 8.30). Because they are of like sign, they repel one-another. The total repulsion of a dislocation by a pile-up is the sum of the repulsions of each dislocation in the pile-up. With *n* dislocations in the pile-up, the stress on the leading dislocation, τ_n, will be

$$\tau_n = n\tau \qquad 8.31$$

It takes a relatively small number in a pile-up to effectively stop further dislocation movement. A pile-up creates a back stress on the source so that the stress to continue to operate the source must rise. This higher stress allows other sources of slightly smaller spacing, *d*, to operate.

CROSS SLIP

A dislocation cannot move on a plane unless the plane contains both the dislocation and its Burgers vector. This requirement uniquely

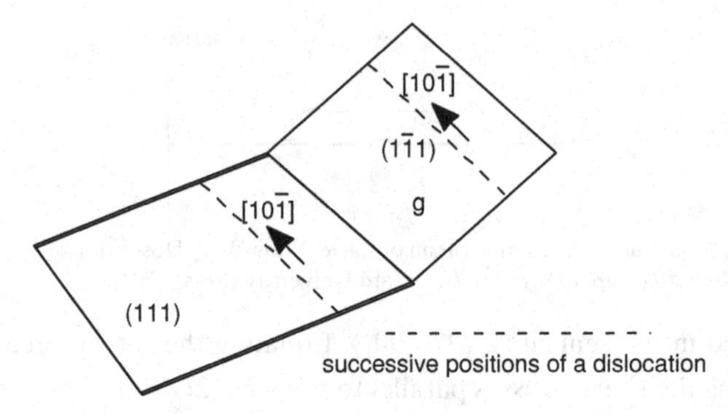

Figure 8.31. A screw dislocation on one slip plane can move onto another plane containing the Burgers vector. This is called *cross slip*.

determines the slip plane, except in the case of screw dislocations. Screw dislocations are parallel to their Burgers vector so they can slip on any plane in which they lie. When a screw dislocation gliding on one plane changes to another plane, it is said to undergo *cross slip*. In principle, screw dislocations should be able to cross slip with ease to avoid obstacles (Figure 8.31).

If, however, a screw dislocation is separated into partials connected by a stacking fault, both partials cannot be screws. They must recombine to form a screw dislocation before they can dissociate on a second plane (Figure 8.32).

Such recombination increases the total energy and this energy must be supplied by the applied stresses aided by thermal activation. How much energy is required depends on the degree of separation of the partials and therefore on the stacking fault energy. If the stacking fault energy is high, the separation of partial dislocations is small so the force required to cause recombination is low. In metals of high stacking fault energy (for example, aluminum), cross slip occurs frequently. In crystals of low stacking fault energy (for example, brass), the separation of partials is large. A high force is required to bring them together so cross slip is rare. This is in accord with observation of slip traces on polished surfaces. Slip traces are very wavy in aluminum and very straight in brass.

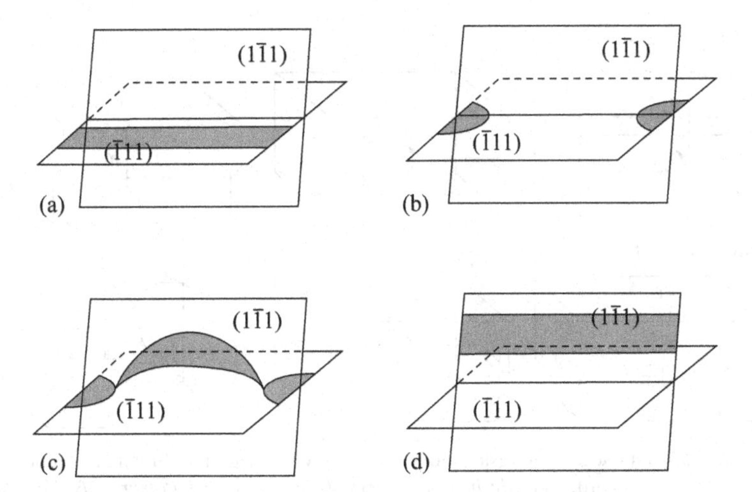

Figure 8.32. Cross slip of a dissociated screw dislocation(a). It must first recombine (b) and then dissociate onto the cross slip plane (c) before it finally can glide on the cross-slip plane (d). From W. F. Hosford, *The Mechanics of Crystals and Textured Polycrystals*, Oxford University Press (1993).

DISLOCATION INTERSECTIONS

The number of dislocations increases as deformation proceeds and the increased number makes their movement become more difficult. This increased difficulty of movement is caused by intersection of dislocations moving on different planes. During easy glide, the rate of work hardening is low because slip occurs on parallel planes and there are few intersections. As soon as slip occurs on more than one set of slip planes, dislocations on different planes will intersect and these intersections impede further motion, causing rapid work hardening.

The nature of dislocations intersections can be understood by considering several types of intersections in simple cubic crystals as illustrated in Figure 8.33. When two dislocations intersect, a jog is created in each dislocation. The direction of the jog is parallel to the Burgers vector of the intersecting dislocation and the length of the jog equals the magnitude of the Burgers vector of the intersecting dislocation.

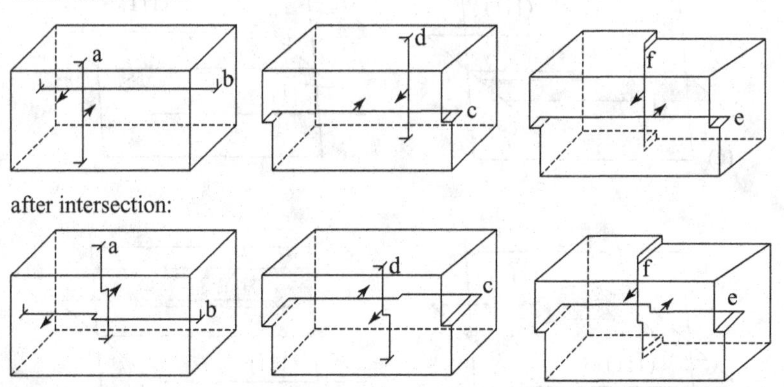

Figure 8.33. Intersection of dislocations. The arrows indicate the direction of motion. From W. F. Hosford, *The Mechanics of Crystals and Textured Polycrystals*, Oxford University Press (1993).

For dislocations a and b in Figure 8.33, the jogs create no problem. The jogs will disappear if the upper part of dislocation a and the right side of dislocation b move slightly faster. The same is true for dislocation c if its left side moves faster. The jog in dislocation d simply represents a ledge in the extra half plane, and it can move with the rest of the dislocation. However, the jogs in dislocations e and f cannot move conservatively. The jogs have an edge character and the direction of motion is not in their slip plane. Figure 8.34 is an enlarged view of dislocations e and f in Figure 8.32. Continued motion of these jogged

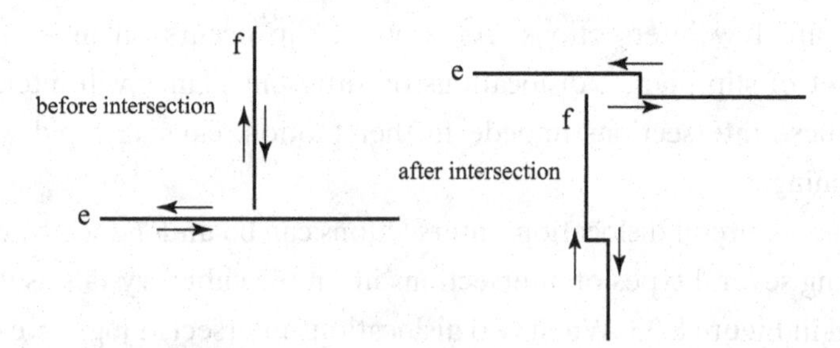

Figure 8.34. Jogs produced in dislocations e and f by their intersection. The arrows indicate the shear caused by continued movement of the dislocations. The jogs are such that they must create interstitial defects as they move. From W. F. Hosford, *The Mechanics of Crystals and Textured Polycrystals*, Oxford University Press (1993).

dislocations would force atoms into interstitial positions. With jogs of the opposite sense, vacancies would be produced.

NOTE OF INTEREST

Erich Schmid (May 4, 1896–October 22, 1983) was born in Bruck-on-the-Mur, Austria and studied in Vienna. His studies were interrupted by service in World War I. He obtained his doctorate in 1920 and remained at the Technische Hochschule in Vienna for two years. In 1922, he left Austria for thirty years to work in various places in Germany and Switzerland. In Berlin, he did research on metal crystals and formulated what is known today as "Schmid's law." During this period, he introduced physical concepts into the study of metals to supplement the earlier chemical concepts. His research with Dehlinger and Sachs forms the basis for much of the present understanding of mechanical behavior of metals. During World War II, Schmid worked on developing and improving substitute materials. In 1951, he returned to the University of Vienna in Austria and remained there until 1967. He served as President of the Austrian Academy of Science for ten years and was responsible for the establishment of several research institutes. He was awarded honorary doctorates by several technical universities in Germany and in 1979 was recognized by Austria's highest achievement award. Scientific organizations and institutes in Germany, Austria, and Japan have given him numerous honors.

From 1933 to 1937, Boas worked in Switzerland, first at the University of Fribourg, and then at the Eidgenössische Technische Hochschule Zurich. While at Fribourg, he and Schmid published their book *Kristallplastizität* (1935), which continued to appear in German (and later English) for the next thirty-five years. Although he was a baptised Lutheran, Boas came from a Jewish family and decided to leave Europe, taking a temporary appointment at the Royal Institution in London, while he searched for a more permanent position. In January 1938, he accepted a two-year appointment at the University of Melbourne, funded by the Carnegie Foundation, in preference to a position at University College, London. For the next nine years, Boas lectured in the metallurgy department of the university, from 1940 as senior lecturer. His book *Introduction to the Physics of Metals and Alloys* (1947) was based on his lectures. His international reputation grew. In 1956, he spent three months in the United States of America as visiting lecturer in metallurgy at Harvard University. He published his third book, *Properties and Structure of Solids* (1971), and was awarded an honorary doctorate of applied science by the university in 1974.

The Frank-Read source was postulated to explain why the number of dislocations in a crystal increased during deformation (rather than decrease as dislocations leave the crystal). F. C. Frank of the University of Bristol and W. T. Read of the Bell Telephone Labs each conceived the idea independently as they traveled to the Symposium on Plastic Deformation of Crystalline Solids at the Carnegie Institute of Technology in Pittsburgh. Each became aware of the other's ideas during informal discussions before the formal conference. In a Pittsburgh pub, they worked out the theory together and decided to make a joint presentation. (F. C. Frank and W. T. Read, *Symposium on Plastic Deformation of Crystalline Solids*, Carnegie Inst. Tech. 1950, v. 44, and F. C. Frank and W. T. Read, Phys. Rev. v. 79 (1950).

REFERENCES

1. E. Schmid, *Proc. Internat. Cong. Appl. Mech.*, Delft (1924).
2. G. I. Taylor, *J. Inst. Metals* v. 62 (1928).
3. G. I. Taylor, *Proc. Roy. Soc.* (London), v. A145 (1934).
4. M. Polanyi, *Z. Physik* v. 89 (1934).
5. E. Orowan, *Z. Physik* v. 89 (1934).
6. J. Burgers, *Proc. Nederlansche Acad. Wettenschappen* v. 42 (1939).

GENERAL REFERENCES

J. P. Hirth and J. Löthe, *Theory of Dislocations*, 2nd Ed., Wiley (1982).

A. H. Cottrell, *Dislocations and Plastic Flow in Crystals*, Oxford University Press (1953).

W. T. Read, *Dislocations in Crystals*, McGraw-Hill (1953).

Weertman and J. R. Weertman, *Elementary Dislocation Theory*, Oxford University Press (1992).
D. Hull and D. J. Bacon, *Introduction to Dislocations*, 3rd Ed., Butterworth Heinemann (1997).
F. McClintock and A. Argon, *Mechanical Behavior of Materials*, Addison-Wesley (1966).

9

TAYLOR AND BISHOP AND HILL MODELS

The first attempts to relate the plastic behavior of polycrystalline metals to the behavior of single crystals were made by Sachs [1] and by Cox and Sopworth [2]. They took the tensile yield strength of the polycrystal to be the average of the yield strengths of the crystals in it. For a single crystal, the tensile yield strength, σ, is given by Schmid's law [3] (equation 9.4). For fcc metals, the average value of $(1/m)$ is 2.238, so according to Sachs' model the tensile yield strength of a randomly oriented polycrystal should be $\sigma = 2.238\tau$.

TAYLOR'S MODEL

Taylor [4, 5] realized that the weakness in Sachs' model is that it assumes that only the most highly slip system is active in each grain, even though more than one system must be active in each grain. To insure that the shape change in each grain of a polycrystal is compatible with neighboring grains, more than one slip system must be active. Taylor analyzed the deformation of a randomly oriented polycrystal of a fcc metal deforming by {111}<110> slip. His analysis was based on finding the amount of slip necessary in each grain to accommodate the required shape change. He assumed that all grains in a polycrystal deform with the same shape change. For an isotropic (randomly oriented) material strained in an x-direction tension test,

$$\varepsilon_y = \varepsilon_z = -(1/2)\varepsilon_x \quad \text{and} \quad \tau_{yz} = \tau_{zx} = \tau_{xy} = 0. \qquad 9.1$$

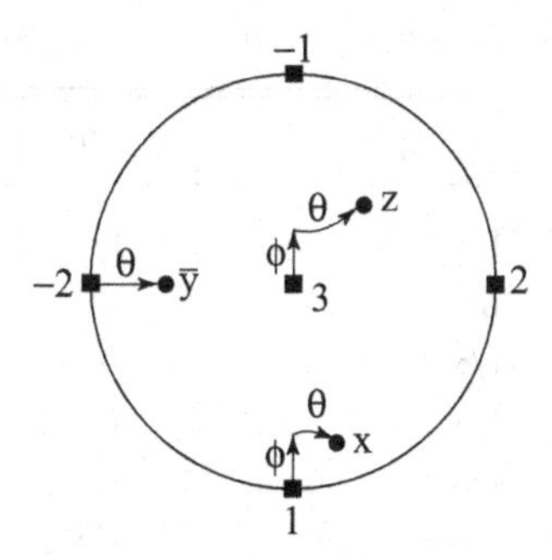

Figure 9.1. Stereographic representation of the relations between the external axes, x, y and z and the cubic crystal axes, 1, 2 and 3. From W. F. Hosford, *The Mechanics of Crystals and Textured Polycrystals*, Oxford Sci. Pub. 1993.

Letting 1, 2 and 3 be the cubic axes of a grain, $\varepsilon_1 + \varepsilon_2 + \varepsilon_3 = 0$, so

$$\varepsilon_3 = -(\varepsilon_1 + \varepsilon_2). \tag{9.2}$$

The external strains, $d\varepsilon_x, d\varepsilon_y \ldots . d\gamma_{xy}$, can be expressed relative to the strains, $d\varepsilon_1, d\varepsilon_2 \ldots . d\gamma_{12}$ along the 1, 2, 3 cubic axes of a crystal and then related to the shear strains on the slip systems. The orientation of the tensile axis, x, relative to the crystal axes, can be described by two angles, θ and ϕ as shown in Figure 9.1. The angles θ and ϕ may be regarded as longitude and latitude. Table 9.1 gives the direction cosines between the two sets of axes.

The strains, $d\varepsilon_1$, $d\varepsilon_2 \ldots . d\gamma_1$, can, in turn, be expressed in terms of the shear strains on the carious slip systems. The {111}<110> slip systems can be identified by the slip planes, a = (111), b = $(\bar{1}\bar{1}1)$, c = $(\bar{1}11)$ and d = $(1\bar{1}1)$ and the slip directions on those planes as shown

Table 9.1. *Direction cosines between the 1, 2, 3 axes and the x, y, z, axes*

	1	2	3
X	$\cos\phi$	$-\sin\theta \sin\phi$	$\cos\theta \sin\phi$
Y	0	$\cos\theta$	$-\sin\theta$
Z	$-\sin\phi$	$\sin\theta \cos\phi$	$\cos\theta \cos\phi$

Table 9.2. *Direction cosines between the cubic axes and the slip elements*

	1 = [100]	2 = [010]	3 = [001]
a = (111)	$1/\sqrt{3}$	$1/\sqrt{3}$	$1/\sqrt{3}$
I = $[01\bar{1}]$	0	$1/\sqrt{2}$	$-1/\sqrt{2}$
II = $[\bar{1}01]$	$-1/\sqrt{2}$	0	$1/\sqrt{2}$
III = $[1\bar{1}0]$	$1/\sqrt{2}$	$-1/\sqrt{2}$	0
b = $(\bar{1}\bar{1}1)$	$-1/\sqrt{3}$	$-1/\sqrt{3}$	$\sqrt{3}$
I = $[0\bar{1}\bar{1}]$	0	$-1/\sqrt{2}$	$-1/\sqrt{2}$
II = [101]	$1/\sqrt{2}$	0	$1/\sqrt{2}$
III = $[\bar{1}10]$	$-1/\sqrt{2}$	$1/\sqrt{2}$	0
c = $(\bar{1}11)$	$-1/\sqrt{3}$	$1/\sqrt{3}$	$1/\sqrt{3}$
I = $[01\bar{1}]$	0	$1/\sqrt{2}$	$-1/\sqrt{2}$
II = [101]	$1/\sqrt{2}$	0	$1/\sqrt{2}$
III = $[\bar{1}\bar{1}0]$	$-1/\sqrt{2}$	$-1/\sqrt{2}$	0
d = $(1\bar{1}1)$	$1/\sqrt{3}$	$-1/\sqrt{3}$	$1/\sqrt{3}$
I = $[0\bar{1}\bar{1}]$	0	$-1/\sqrt{2}$	$-1/\sqrt{2}$
II = $[\bar{1}01]$	$-1/\sqrt{2}$	0	$1/\sqrt{2}$
III = [110]	$1/\sqrt{2}$	$1/\sqrt{2}$	0

in Figure 8.9. Table 9.2 gives the direction cosines between the cubic axes and the slip plane and directions.

Using Table 9.2, the strains relative to the cubic axes can be expressed as:

$$
\begin{aligned}
d\varepsilon_1 &= (-d\gamma_{\mathrm{aII}} + d\gamma_{\mathrm{aIII}} - d\gamma_{\mathrm{bII}} + d\gamma_{\mathrm{bIII}} - d\gamma_{\mathrm{cII}} + d\gamma_{\mathrm{cIII}} - d\gamma_{\mathrm{dII}} + d\gamma_{\mathrm{dIII}})/\sqrt{6}\\
d\varepsilon_2 &= (d\gamma_{\mathrm{aI}} - d\gamma_{\mathrm{aIII}} + d\gamma_{\mathrm{bI}} - d\gamma_{\mathrm{bIII}} + d\gamma_{\mathrm{cI}} - d\gamma_{\mathrm{cIII}} + d\gamma_{\mathrm{dI}} - d\gamma_{\mathrm{dIII}})/\sqrt{6}\\
d\varepsilon_3 &= (-d\gamma_{\mathrm{aI}} + d\gamma_{\mathrm{aII}} - d\gamma_{\mathrm{bI}} + d\gamma_{\mathrm{bII}} - d\gamma_{\mathrm{cI}} + d\gamma_{\mathrm{cII}} - d\gamma_{\mathrm{dI}} + d\gamma_{\mathrm{dI}})/\sqrt{6}\\
d\gamma_{23} &= (d\gamma_{\mathrm{aII}} - d\gamma_{\mathrm{aIII}} - d\gamma_{\mathrm{bII}} - d\gamma_{\mathrm{bIII}} + d\gamma_{\mathrm{cII}} - d\gamma_{\mathrm{cIII}} - d\gamma_{\mathrm{dII}} + d\gamma_{\mathrm{dIII}})/\sqrt{6}\\
d\gamma_{31} &= (-d\gamma_{\mathrm{aI}} + d\gamma_{\mathrm{aIII}} + d\gamma_{\mathrm{bII}} - d\gamma_{\mathrm{bIII}} + d\gamma_{\mathrm{cI}} - d\gamma_{\mathrm{cIII}} - d\gamma_{\mathrm{dI}} + d\gamma_{\mathrm{dIII}})/\sqrt{6}\\
d\gamma_{12} &= (d\gamma_{\mathrm{aI}} - d\gamma_{\mathrm{aII}} + d\gamma_{\mathrm{bI}} - d\gamma_{\mathrm{bII}} - d\gamma_{\mathrm{cI}} + d\gamma_{\mathrm{cII}} - d\gamma_{\mathrm{dI}} + d\gamma_{\mathrm{dII}})/\sqrt{6}
\end{aligned}
\qquad 9.3
$$

With $\varepsilon_3 = -\varepsilon_2 - \varepsilon_1$, there are five independent strains. Taylor reasoned that to produce these five independent strains, at least five independent slip systems are required. Not every combination of five

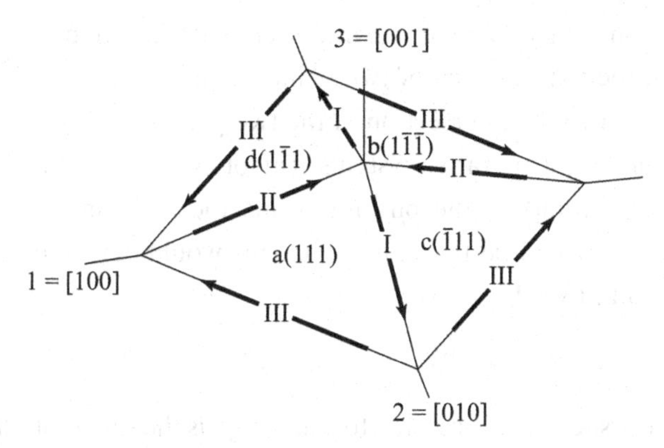

Figure 9.2. Half of an octahedron representing the {111}<110> slip systems. From W. F. Hosford, *The Mechanics of Crystals and Textured Polycrystals*, Oxford Sci. Pub. 1993.

slip systems can be used however. For an arbitrary shape change, the five systems must be independent in the sense that the shear on any one cannot be produced by a combination of shears on the other four. Figure 9.3 gives examples of combinations that are not independent. Three slip directions on a plane (9.3A), three plane sharing a common slip direction (9.3B), two directions on each of two planes with the fifth

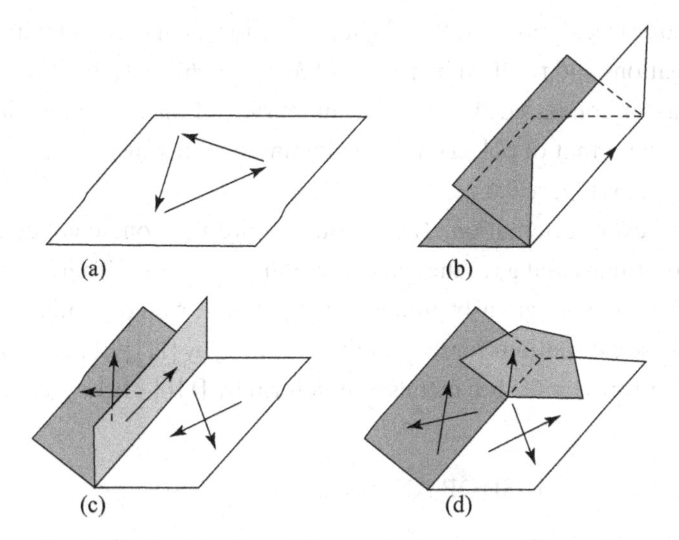

Figure 9.3. Combinations of slip systems that are not independent.

slip direction lying parallel to the intersection of the two planes (9.3C) and two directions on each of two planes with the fifth slip direction perpendicular to the intersection of the two planes (9.3D).

To identify the operative systems, Taylor assumed that the correct combination would be the one for which the total amount of slip, $d\gamma = \Sigma|d\gamma_i|$, would be the minimum. This would correspond to the least amount of work

$$dW = \tau d\gamma = \sigma_x d\varepsilon_x, \tag{9.4}$$

where τ, the shear stress needed to cause slip, is the same for all active slip systems. The relative strength of any grain is therefore,

$$M = \sigma_x/\tau = d\gamma/d\varepsilon_x = dW/(\tau d\varepsilon_x), \tag{9.5}$$

where M is called the Taylor factor. He found the minimum value of M for each orientation of grain by making calculations for many combinations of slip systems. Taylor found an average of $M_{av} = 3.06$. In his original work, he mistakenly omitted many combinations of slip systems, but later work showed that his value of 3.06 is nearly correct. Figure 9.4 is his plot of M over the orientation triangle.

Later calculations [7, 8], (Figure 9.5) showed agreement with his calculations and resulted in a value of $M_{av} = 3.067 \pm 0.001$.

Taylor compared the stress strain curves of an aluminum single crystal with that of polycrystalline aluminum by taking $\sigma = 3.06\tau$ and $\varepsilon = \gamma/3.06$ (Figure 9.6.)

For every orientation, Taylor found more than one combination of slip systems that gave the same minimum value of M. Therefore, he could not unambiguously predict how the lattice would rotate.

These calculations for fcc metals deforming by {111}<110> slip are also appropriate for bcc metals that deform by {110}<111> slip.

BISHOP AND HILL ANALYSIS

Bishop and Hill [7, 8] used a different approach. Instead of finding the combination of slip systems that corresponded to the minimum work,

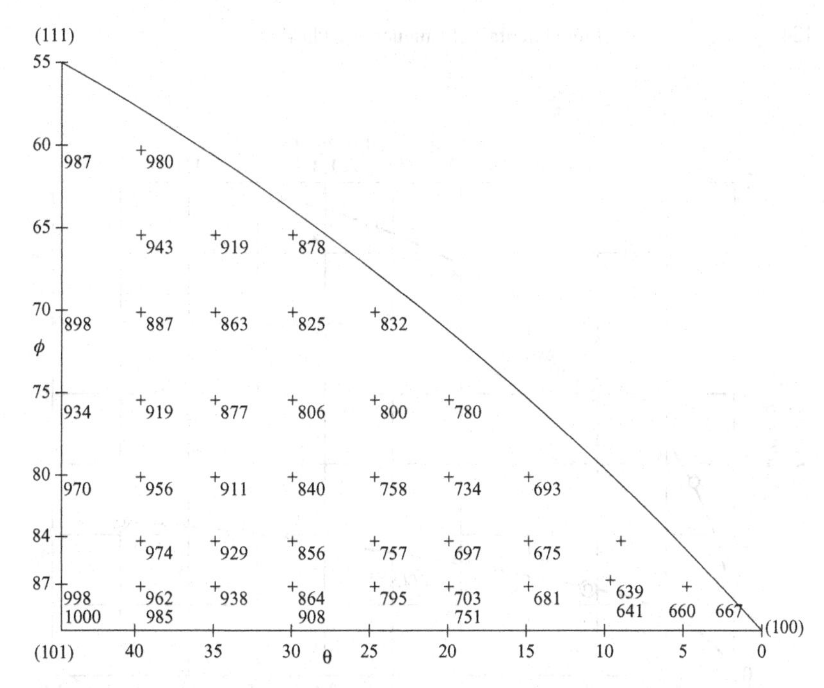

Figure 9.4. Taylor's plot of the orientation dependence of M for axially symmetric flow. The M-values equal the plotted numbers multiplied by $(3/2)\sqrt{6}/1000$. From G. I Taylor, *J. Inst. Met.* 62 (1938).

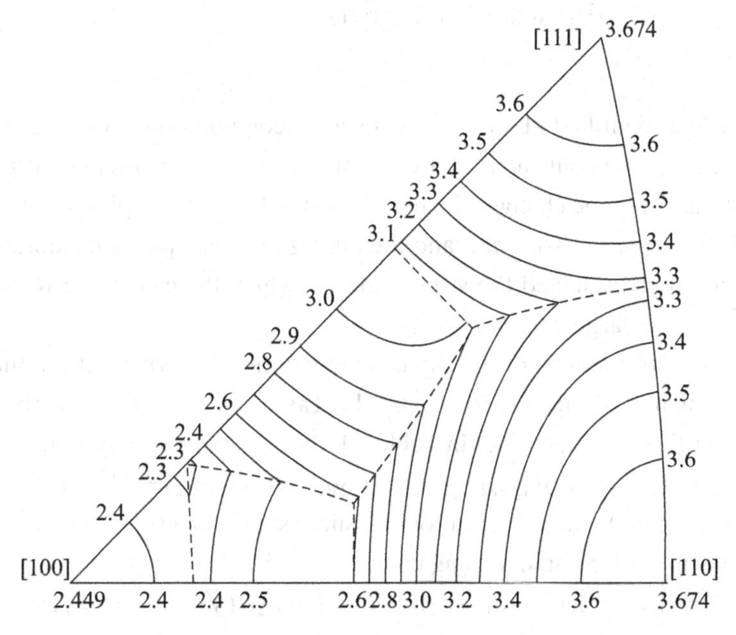

Figure 9.5. Orientation dependence of M for axially symmetric flow. From G. Y. Chin and W. L. Mammel, *Trans. TMS-AIME* 239 (1967).

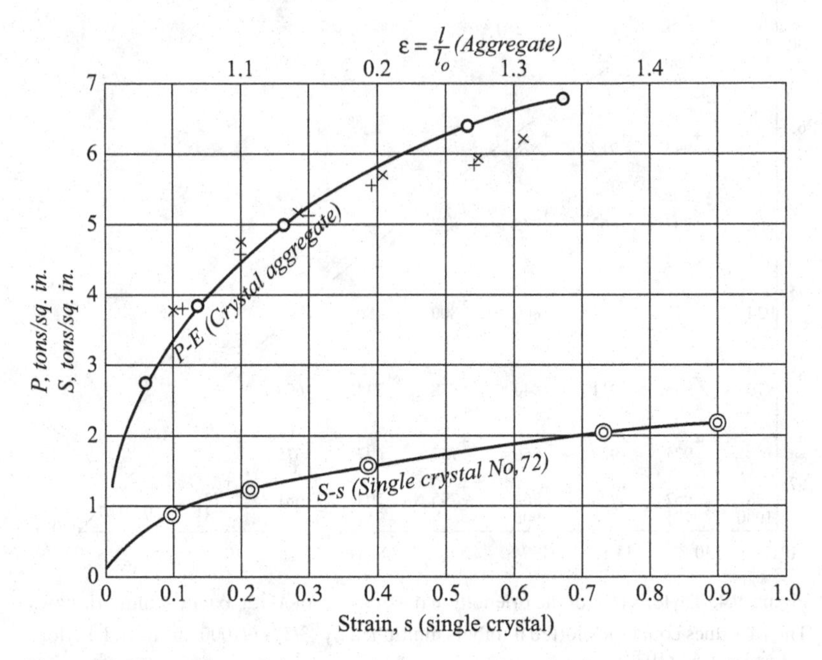

Figure 9.6. Comparison of the stress strain curves of an aluminum polycrystal and an aluminum single crystal. The polycrystal curve was calculated by taking $\sigma_x = M\tau$ and $\varepsilon_x = \gamma_d/M$. From G. I Taylor, *J. Inst. Met.* 62 (1938).

they first identified all of the stress states (combinations of $\sigma_1, \sigma_2, \sigma_3, \sigma_{23,31}, \sigma_{12}$) that could activate five or more slip systems simultaneously. For a given shape change with $\sigma_3 = 0$, they calculated the plastic work, $dw/d\varepsilon_x = \sigma_1/d\varepsilon_x + \sigma_2/d\varepsilon_x$, and then using the principle of maximum virtual work, selected the stress state for which the calculated work, $dw/d\varepsilon_x$ was largest.

The designations of the slip elements in this development are the same as used earlier in describing the Taylor analysis. For example, using the direction cosines in Table 6.1, the shear stress on system aI is $\sigma_{a1} = \ell_{aI}\ell_{11}\sigma_{11} + \ell_{a2}\ell_{12}\sigma_{22} + \cdots (\ell_{a1}\ell_{12}\sigma_{11} + \ell_{a2}\,\ell_{11}\sigma_{12}) = (1/\sqrt{6})(\sigma_2 - \sigma_3 - \sigma_{31} + \sigma_{12})$. Slip will occur on this slip system when $(\sigma_2 - \sigma_3 - \sigma_{31} + \sigma_{12})/\sqrt{6} = \pm 1$. Similarly using the notation $A = (\sigma_2 - \sigma_3)/(\sqrt{6}\tau)$, $B = (\sigma_3 - \sigma_{31})/(\sqrt{6}\tau)$, $C = (\sigma_1 - \sigma_2)/(\sqrt{6}\tau)$, $F = \sigma_{23}/(\sqrt{6}\tau)$, $G = \sigma_{31}/(\sqrt{6}\tau)$

Table 9.3. *Stress states to activate slip on the twelve slip systems*

System	Stress state
aI	A − G + H = ±1
bI	A + G + H = ±1
cI	A + G − H = ±1
dI	A − G − H = ±1
aII	B + F − H = ±1
bII	B − F − H = ±1
cII	B + F + H = ±1
dII	B − F + H = ±1
aIII	C − F + G = ±1
bIII	C + F − G = ±1
cIII	C − F − G = ±1
dIII	C + F + G = ±1

and $H = \sigma_{12}/(\sqrt{6}\tau)$, the yield criteria for all twelve slip systems can be written as shown in Table 9.3.

Thus, the shear stress for slip is reached when one of these stress states is reached without exceeding any other. In other words,

$$|A \pm G \pm H| \leq 1, \quad |B \pm F \pm H| \leq 1, \quad |C \pm F \pm G| = 1. \qquad 9.7$$

This results in twenty-eight stress states (combinations of *A, B,*... *H*) that can satisfy five or more slip systems. These are listed in Table 9.4, together with the slip systems activated.

Table 9.4 shows that twelve of the stress states can activate eight slip systems simultaneously and that the other sixteen stress states can activate six slip systems simultaneously. This explains why Taylor found that more that one combination of five slip systems gave the same value of *M*. Bishop and Hill used the principle of maximum virtual work to determine which stress state is appropriate for a given shape change.

Table 9.4. *Bishop and Hill stress states and corresponding slip systems*

	Stress state						Slip systems											
	A	*B*	*C*	*F*	*G*	*H*	aI	aII	aIII	bI	bII	bIII	cI	cII	cIII	dI	dII	dIII
1	1	−1	0	0	0	0	+	−	0	+	−	0	+	−	0	+	−	0
2	0	1	−1	0	0	0	0	+	−	0	+	−	0	+	−	0	+	−
3	−1	0	1	0	0	0	−	0	+	−	0	+	−	0	+	−	0	+
4	0	0	0	1	0	0	0	+	−	0	−	+	0	+	−	0	−	+
5	0	0	0	0	1	0	−	0	+	+	0	−	+	0	−	−	0	+
6	0	0	0	0	0	1	+	−	0	+	−	0	−	0	+	−	0	+
7	1/2	−1	1/2	0	1/2	0	0	−	+	+	−	0	+	−	0	0	−	+
8	1/2	−1	1/2	0	−1/2	0	+	−	0	0	−	+	0	−	+	+	−	0
9	−1	1/2	1/2	1/2	0	0	−	+	0	−	0	+	−	0	+	−	+	0
10	−1	1/2	1/2	−1/2	0	0	−	0	+	−	+	0	−	0	+	−	+	0
11	1/2	1/2	−1	0	0	1/2	+	0	−	+	0	−	0	+	−	0	+	−
12	1/2	1/2	−1	0	0	−1/2	0	+	−	0	+	−	+	0	−	+	0	−
13	1/2	0	−1/2	1/2	0	1/2	+	0	−	+	−	0	0	+	−	0	0	0
14	1/2	0	−1/2	−1/2	0	1/2	+	−	0	+	0	−	0	0	0	0	+	−
15	1/2	0	−1/2	1/2	0	−1/2	0	+	−	0	0	0	+	0	−	+	−	0
16	1/2	0	−1/2	−1/2	0	−1/2	0	0	0	0	+	−	+	−	0	+	−	0
17	0	−1/2	1/2	0	1/2	1/2	0	−	+	+	−	0	0	0	0	−	0	+
18	0	−1/2	1/2	0	−1/2	1/2	+	−	0	0	−	+	−	0	+	0	0	0
19	0	−1/2	1/2	0	1/2	−1/2	−	0	+	0	0	0	+	−	0	0	−	+
20	0	−1/2	1/2	0	−1/2	−1/2	0	0	0	−	0	+	0	−	+	+	−	0
21	−1/2	1/2	0	1/2	1/2	0	−	+	0	0	0	0	0	+	−	−	+	0
22	−1/2	1/2	0	−1/2	1/2	0	−	0	+	0	+	−	0	0	0	−	+	0
23	−1/2	1/2	0	1/2	−1/2	0	0	+	−	−	0	+	−	+	0	0	0	0
24	−1/2	1/2	0	−1/2	−1/2	0	0	0	0	−	+	0	−	0	+	0	+	−
25	0	0	0	1/2	1/2	−1/2	−	+	0	0	0	0	+	0	−	0	−	+
26	0	0	0	1/2	−1/2	1/2	+	−	0	0	−	+	−	+	0	0	0	0
27	0	0	0	−1/2	1/2	1/2	0	−	+	+	0	−	0	0	0	−	+	0
28	0	0	0	1/2	1/2	1/2	0	0	0	+	−	0	0	+	−	−	+	0

The five stress states for axially symmetric flow are listed in Table 9.5. Figure 9.7 shows the orientations for which each is appropriate.

YIELD LOCUS CALCULATIONS

The models of Taylor and of Bishop and Hill models have been extended to predict plastic behavior under stress states other than uniaxial tension by assuming various shape changes [6]. Figure 9.8 is

Table 9.5. *Yield locus for randomly oriented fcc metals strain path*

$\rho = \varepsilon_y/\varepsilon_z$	σ_x/τ	σ_y/τ
0.500	3.067 ± 0.001	0.000
0.375	3.156 ± 0.004	0.207 ± 0.006
0.250	3.228 ± 0.004	0.439 ± 0.011
0.125	3.284 ± 0.002	0.748 ± 0.011
0.050	3.313 ± 0.006	1.080 ± 0.036
0.000	3.323 ± 0.002	1.651 ± 0.021

a stereographic representation showing that how any orientation can be described by three angles. θ, ϕ , and α.

For every shape change all grains are assumed to deform with the same shape change. Isotropic was approximated with a large number of randomly oriented grains. To assure randomness, θ can be chosen randomly between 0° and 45°, α randomly between 0° and 180° and sinϕ randomly between 0 and 1. This covers three basic orientation triangles. Sinϕ was randomly chosen between 0 and 1 because the spherical area for a given value of $\Delta\phi$ is greater near the equator

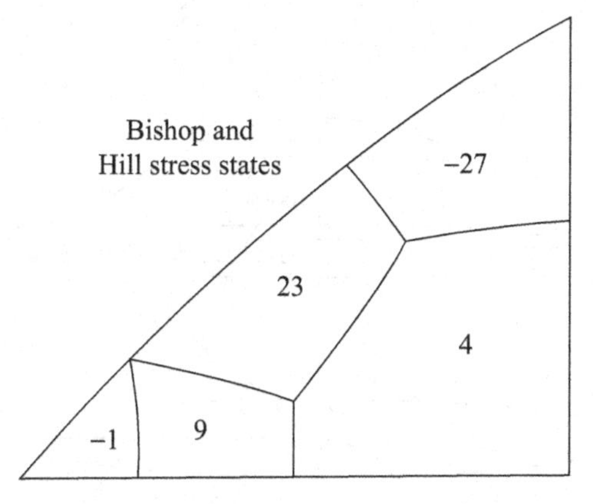

Figure 9.7. Basic orientation triangle showing the appropriate Bishop and Hill stress stats for axially symmetric deformation. From W. F. Hosford, *The Mechanics of Crystals and Textured Polycrystals*, Oxford Sci. Pub. 1993.

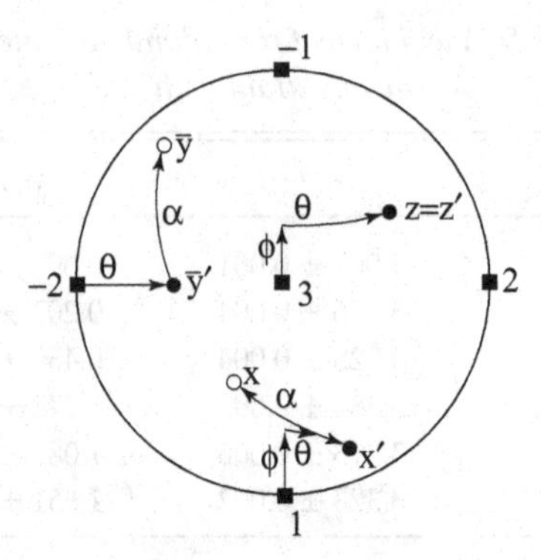

Figure 9.8. Stereographic representation of the angles between the external axes, x, y, z and the cubic axes 1, 2, 3. From W. F. Hosford, *The Mechanics of Crystals and Textured Polycrystals*, Oxford Sci. Pub. 1993.

($\phi = 0$) than near the pole ($\phi = \pi/2$) as shown in Figure 9.9. The circumference at ϕ is $2\pi\cos\phi d\phi$ so the spherical area between ϕ_1 and ϕ_2 equals $\sin\phi_2 - \sin\phi_1$.

Table 9.5 gives the coordinates of the upper-bound yield locus calculated for fcc metals deforming by {111}<110> slip and Figure 9.10 is a plot of the yield locus together with the predictions of the Tresca, von Mises criteria, and equation 2.15 with $a = 8$.

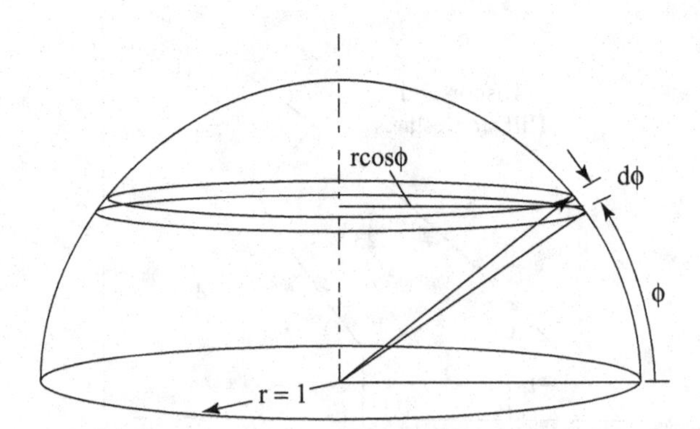

Figure 9.9. Hemisphere of unit radius. The area between ϕ and $\phi + \Delta\phi = 2\pi\cos\phi\, d\phi$. From W. F. Hosford, *The Mechanics of Crystals and Textured Polycrystals*, Oxford Sci. Pub. 1993.

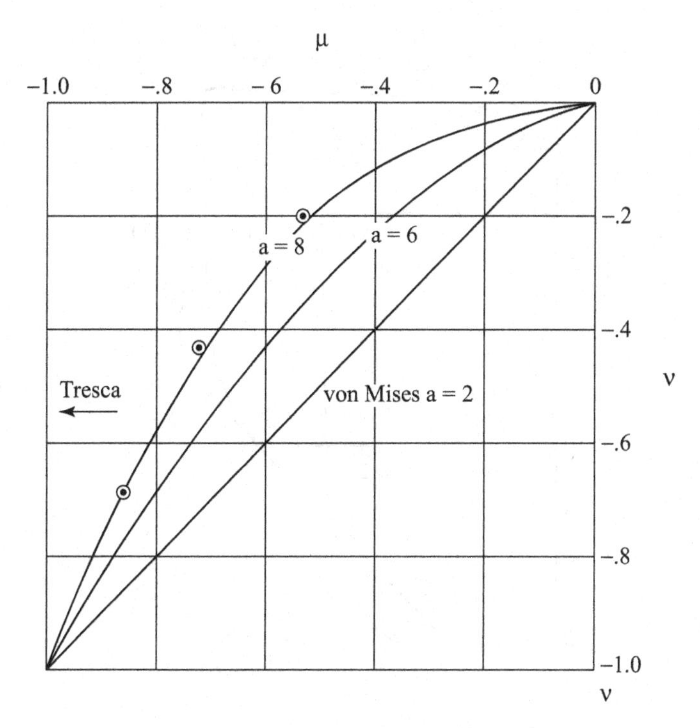

Figure 9.10. Plot of the yield locus together with the predictions of the Tresca, von Mises criteria and equation 1.15 with $a = 6$ and 8.

FIBER TEXTURES

The analyses of Taylor and of Bishop and Hill can be extended to predict the anisotropy of plasticity. Fiber textures are rotationally symmetric about the fiber axis. Figure 9.5 can be used to predict the relative strengths of different fiber textures. For <111> and <110> $M = 1.5\sqrt{6}$, so wires with these textures are much stronger than wires with a <100> texture ($M = \sqrt{6}$).

SHEET TEXTURES

To calculate yield loci of sheet textures various sheet textures were assumed. For each, rotational symmetry was simulated by components rotated by increments of $\Delta\alpha = 5°$ about the normal. Various shape

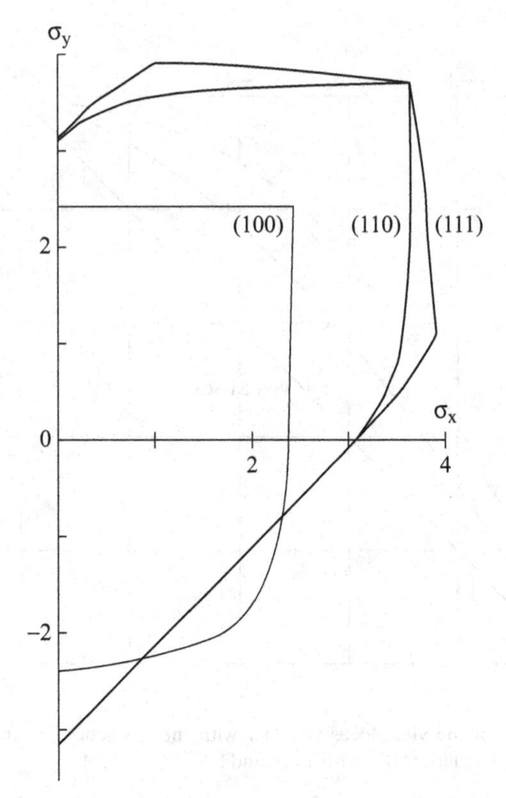

Figure 9.11. Yield loci for sheets with textures that are rotationally symmetric about <100>, <111> and <110>. From W. F. Hosford, *The Mechanics of Crystals and Textured Polycrystals*, Oxford Sci. Pub. 1993.

changes were assumed and the stresses for all orientations were calculated and averaged. Figure 9.11 shows the results of three such calculations for sheets with textures that are rotationally symmetric about <100>, <111>, and <110>. In each case, rotational symmetry was approximated by rotating α by 5° increments about the normal.

Yield loci for more complicated textures were calculated by assuming sheets with five crystallographic directions normal to the sheet. The fraction of grains with each normal was randomly chosen. Rotational symmetry was simulated by rotational increments of $\Delta\alpha =$ 5° increments about the normal. Figure 9.12 shows the results of such

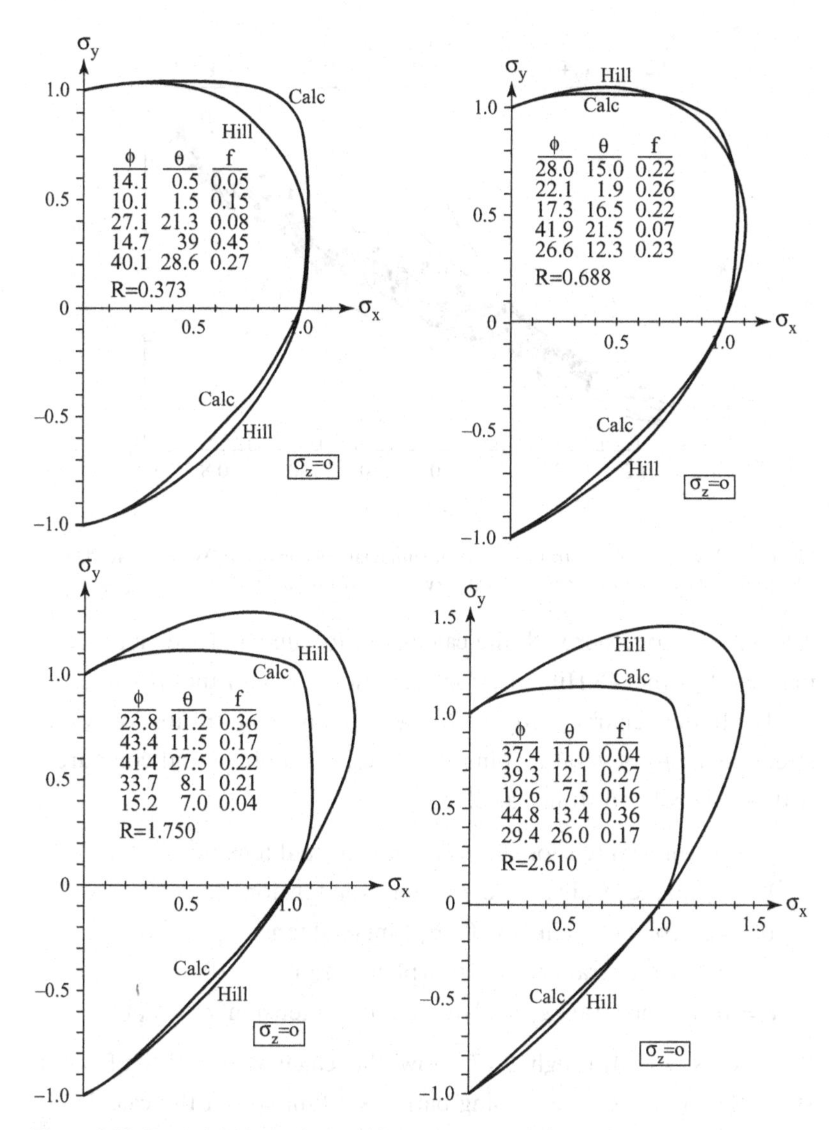

Figure 9.12. Four yield loci calculated for textures consisting of five randomly oriented sheet normals and rotational symmetry about the normals. Also shown are the predictions of Hill's 1948 criterion (equation 4.24) for the calculated R-value. From W. F. Hosford, *The Mechanics of Crystals and Textured Polycrystals*, Oxford Sci. Pub. 1993.

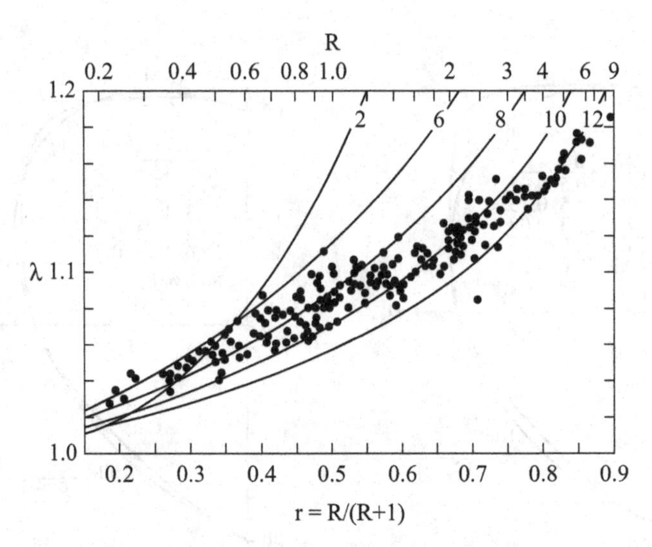

Figure 9.13. $\lambda = \sigma_x$ (plane strain $\varepsilon_y = 0)/\sigma_x$ (uniaxial tension) vs. R. W. F. Hosford, *The Mechanics of Crystals and Textured Polycrystals*, Oxford Sci. Pub. 1993.

calculations, together with the calculated R-value and the yield locus predicted by the '48 Hill criterion (equation 7.24) for that R-value.

To better characterize such calculations, the ratios of yield strengths along several loading paths were examined and compared with the calculated strain ratio, R.

$\chi = \sigma_x$ (biaxial tension, $\sigma_x = \sigma_y)/\sigma_x$ (uniaxial tension)
$\lambda = \sigma_x$ (plane strain $\varepsilon_y = 0)/\sigma_x$ (uniaxial tension)
$\psi = 2\sigma_x$ (plane strain $\varepsilon_z = 0)/\sigma_x$ (uniaxial tension)
$\beta = \sigma_x$ (plane strain $\varepsilon_y = 0)/\sigma_x$ (plane strain $\varepsilon_z = 0$)
$\xi = \sigma_x$ (plane strain $\varepsilon_y = 0)/\sigma_x\ \sigma_x$ (biaxial tension, $\sigma_x = \sigma_y$)

Figures 9.13 through 9.17 show the calculated ratios of yield strengths under several loading paths as a function of the calculated R-value, along with predictions of equation 7.30.

Examination of these figures reveals that there is no unique value of a, but that the trends are best approximated by $a = 8$ in equation 7.30. They also indicate the possibility of the anomaly of χ being greater than one with of R being less than one.

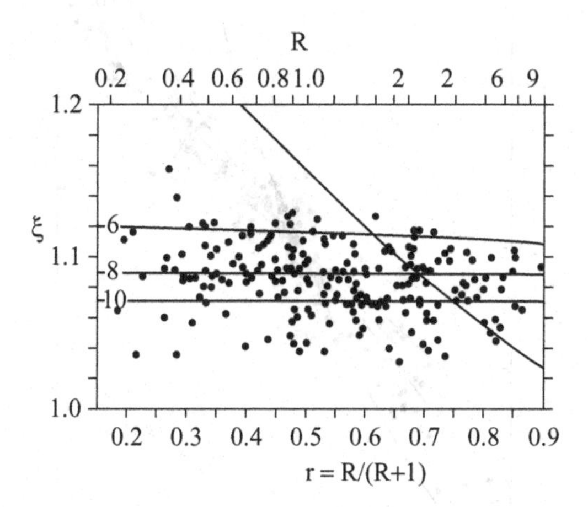

Figure 9.14. $\xi = \sigma_x$ (plane strain $\varepsilon_y = 0)/\sigma_x$ (biaxial tension, $\sigma_x = \sigma_y$) vs R. From W. F. Hosford, *The Mechanics of Crystals and Textured Polycrystals*, Oxford Sci. Pub. 1993.

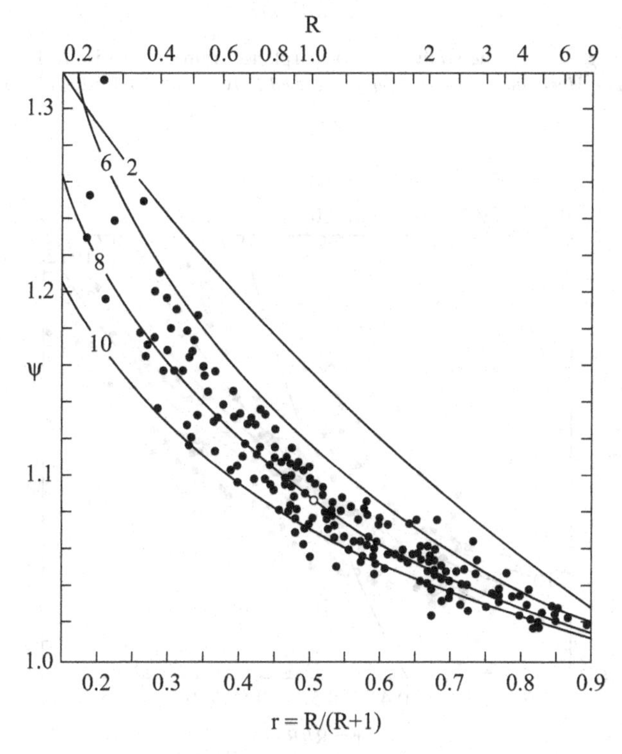

Figure 9.15. $\psi = 2\sigma_x$ (plane strain $\varepsilon_z = 0)/\sigma_x$ (uniaxial tension) vs. R. From W. F. Hosford, *The Mechanics of Crystals and Textured Polycrystals*, Oxford Sci. Pub. 1993.

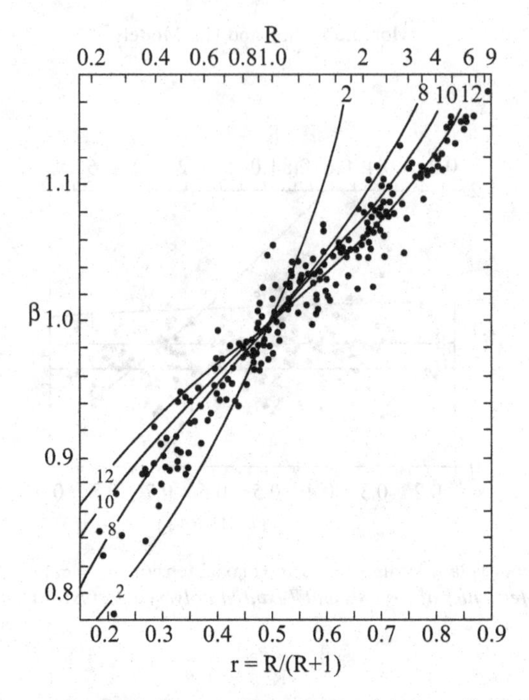

Figure 9.16. $\beta = \sigma_x$ (plane strain $\varepsilon_y = 0)/\sigma_x$ (plane strain $\varepsilon_z = 0$) vs. R. From W. F. Hosford, *The Mechanics of Crystals and Textured Polycrystals*, Oxford Sci. Pub. 1993.

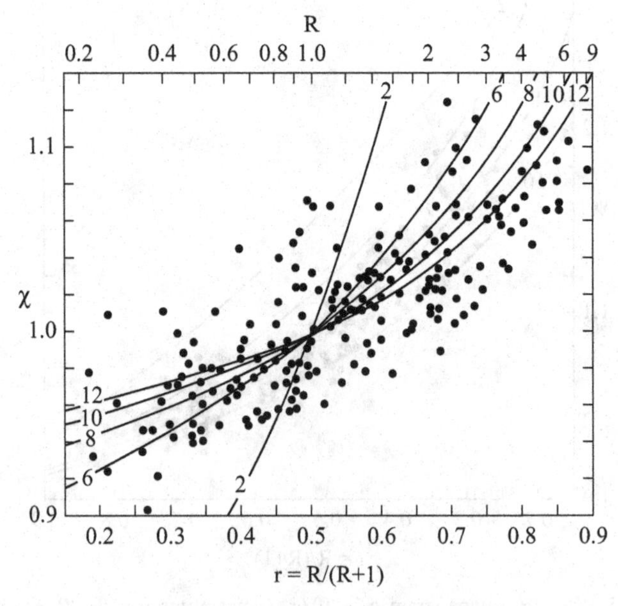

Figure 9.17. $\chi = \sigma_x$ (biaxial tension, $\sigma_x = \sigma_y)/\sigma_x$ (uniaxial tension) vs. R. From W. F. Hosford, *The Mechanics of Crystals and Textured Polycrystals*, Oxford Sci. Pub. 1993.

The Taylor and the Bishop and Hill upper-bound models are entirely equivalent. While they were initially formulated for {111}<110> slip in fcc metals, they also apply to bcc metals that deform by {110}<111> slip. They involve several critical assumptions. Kocks [11] gives an excellent discussion of these. One of these is the tendency of grains to depart from the assumption that all grains deform homogeneously and with the same strain as the whole polycrystal. Another is the assumption that strain hardening depends only on $\Sigma[d\gamma_i]$.

CALCULATIONS INCLUDING STRAIN HARDENING

The effects of strain hardening can be included in upper-bound calculations by assuming

$$\tau_i = K\gamma_i^n \quad 9.7$$

The work per volume in any grain is then $w_i = K\gamma_i{}^{n+1}/(n+1)$. Substituting $\gamma_i = M_i\varepsilon_x$,

$$w_i = KM_i^{n+1}\varepsilon_x^{n+1}/(n+1). \quad 9.8$$

For a polycrystals consisting of g grains,

$$w = (1/g)\sum_i w_i = K\varepsilon_x^{n+1}\sum_i M_i^{n+1}/[g(n+1)], \quad 9.9$$

where the summation is over all grains. Yielding can be defined to occur when w reaches a critical value, w^*. Then,

$$\varepsilon_x = \left(\frac{w^*(g/K)(n+1)}{\sum_i M^{n+1}}\right) \quad 9.10$$

The values of σ_x/τ and σ_y/τ for plane stress, $\sigma_z = 0$ can be found for every shape change.

$$\sigma_{xi} = (\sigma_x/\tau)_i\tau_i = (\sigma_x/\tau)KM_i^n\varepsilon_x^n = (\sigma_x/\tau)KM_i^n\left(\frac{(g/K)(n+1)w^*}{\sum_i M_i^{n+1}}\right)^{n/(n+1)}$$

9.11

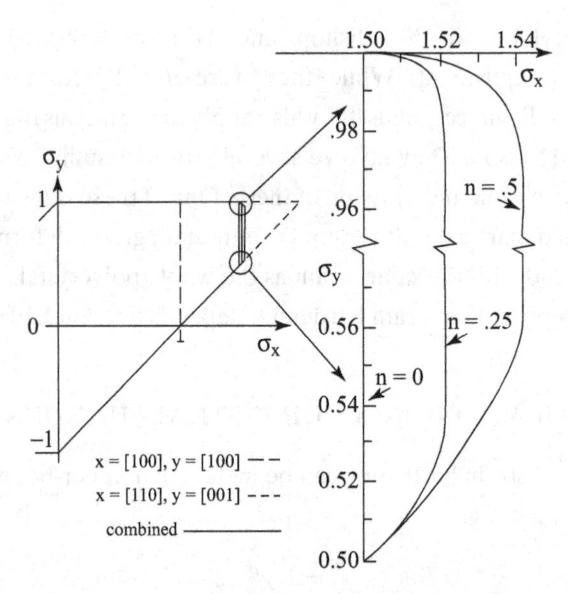

Figure 9.18. Inclusion of strain hardening rounds the corners of yield loci. From W. F. Hosford, "Incorporating Work Hardening Effects in Yield Loci Calculations," in *Proceedings of the 5th International Conference on the Strength of Metals and Alloys*. v. 2, Pergamon Press, 1979.

The average value of σ is

$$\sigma_x = A \sum_i \left(\frac{(\sigma_x/\tau)_i M_i^n}{\left(\sum_i M_i^{n+1}\right)^{n/(n+1)}} \right) \qquad 9.12$$

where A equal $(K/g)^{1/n+1}[w^*(n+1)]^{n/(n+1)}]$ for all grains. Similarly,

$$\sigma_y = A \sum_i \left(\frac{(\sigma_y/\tau)_i M_i^n}{\left(\sum_i M_i^{n+1}\right)^{n/(n+1)}} \right) \qquad 9.13$$

The locus shape is independent of w^*. Figure 9.18 shows how incorporating work hardening changes the shape of the yield locus for a mixture of two orientations by rounding corners.

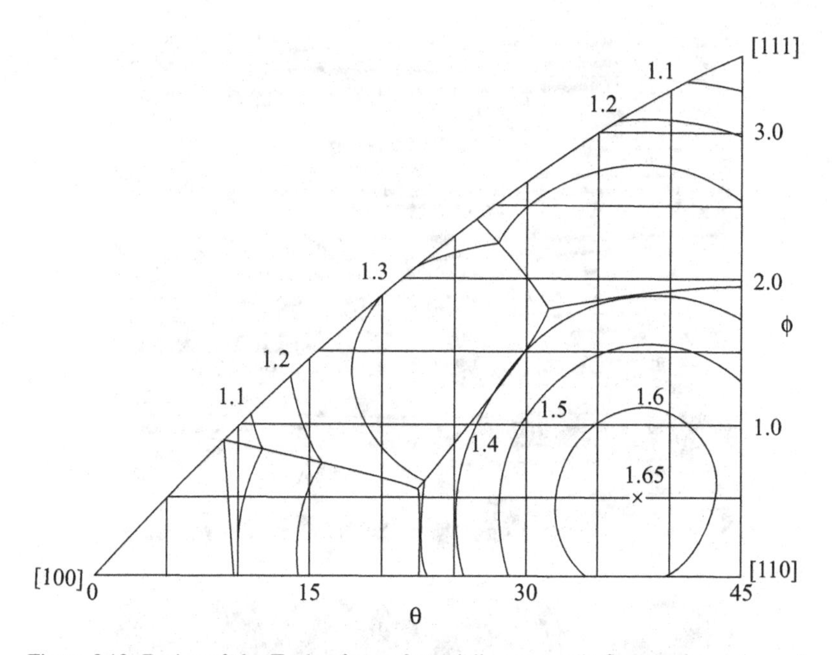

Figure 9.19. Ratios of the Taylor factor for axially symmetric flow to the reciprocal of the Schmid factor. From W. F. Hosford, *The Mechanics of Crystals and Textured Polycrystals*, Oxford Sci. Pub. 1993.

RATE-SENSITIVE MODEL

A rate-sensitive model based on the Taylor assumptions was proposed by Aasaro and Needleman [13]. Each of the six or eight slip systems is assumed to be active and the shear stress on it is assumed to be $\tau = c\gamma^m$. Even with very low levels of m, the ambiguity of how much slip occurs on each system vanishes allowing unambiguous predictions of lattice rotation.

DEVIATIONS FROM OVERALL SHAPE CHANGE

The tendency of an individual grain to deform with a shape change deviating from that of the polycrystal depends on its orientation. It depends on the difference between the Taylor factor and the reciprocal of the Schmid factor for axially symmetric flow. Figure 9.19 shows

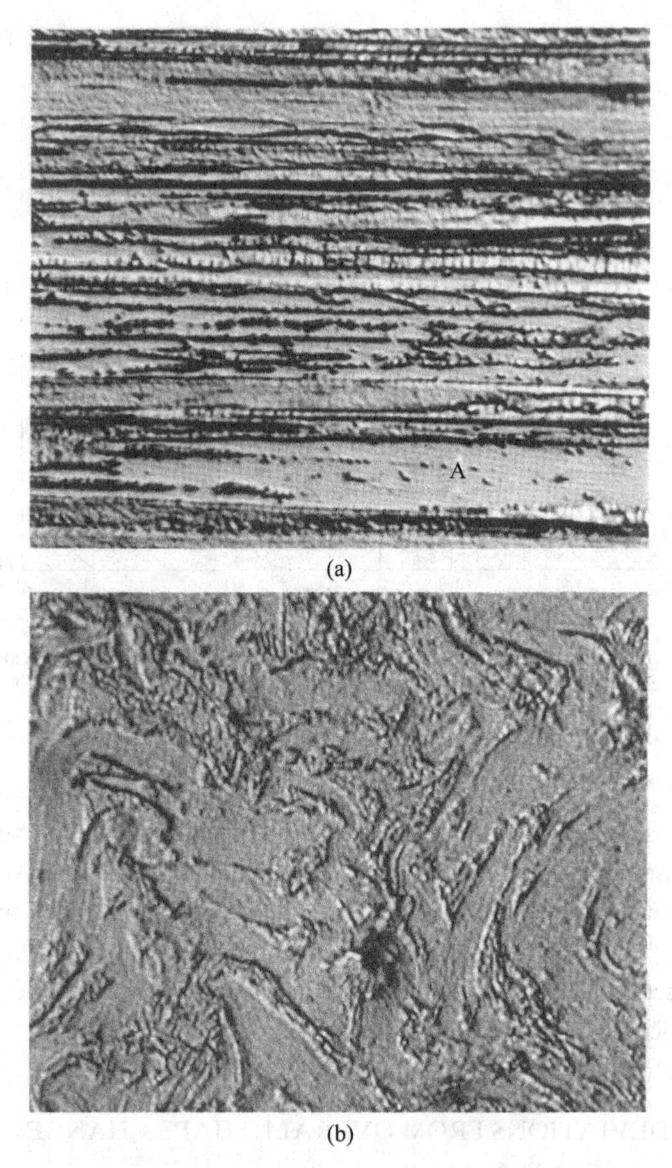

Figure 9.20. Microstructure of heavily drawn tungsten wire. Longitudinal section (a) and transverse section (b). From Peck and Thomas. From J. F. Peck and D. A. Thomas *TMS-AIME* v 221 (1961).

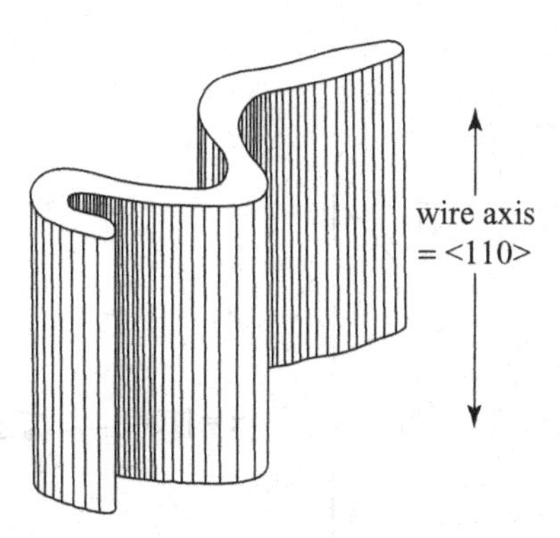

Figure 9.21. The grain shape in drawn bcc metals are like ribbons folded about the wire axis. From W. F. Hosford, *The Mechanics of Crystals and Textured Polycrystals*, Oxford Sci. Pub. 1993.

the ratio of the Taylor factor to the reciprocal of the Schmid factor. The Taylor factor and the reciprocal of the Schmid factors are identical for [100] and [111] so these orientations are not likely to deviate from axially symmetric flow. Orientations near [110] are most likely to deviate from axially symmetric flow.

Wire drawing of bcc metals develops strong <110> fiber textures. Figure 9.20 shows electron micrographs of such wire taken by Peck and Thomas. [14] These suggest that the shape of the grains in heavily drawn tungsten wire are shaped like ribbons curled by bending about the wire axis (Figure 9.21). Such observations have been explained [15] as a result of the difficulty in achieving axially symmetric flow about a <110> axis. As shown in Figure 9.22, two of the four <111> slip directions are perpendicular to the fiber axis so slip in these directions make no contribution to the elongation. The Taylor factor for axially symmetric for is 1.5 times that for plane strain ($\varepsilon_y = 0$) so there is a strong propensity for plane strain. Compatibility with neighboring grains can be maintained if the grains curl about one another. Van Houtte [16] made calculations that included the work of curling. After a strain of about 2, the work for deforming by plane strain is about 0.7 time that for axially symmetric flow.

An analogous microstructure develops in fcc metals after severe compression because of the <110> compression texture [15].

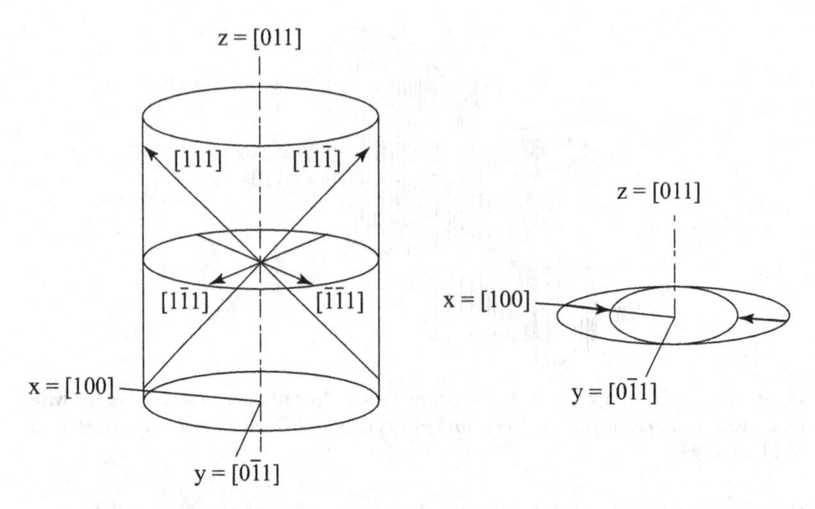

Figure 9.22. Arrangement of the <111> slip directions in a bcc crystal oriented for extension parallel to [011]. From W. F. Hosford, *The Mechanics of Crystals and Textured Polycrystals*, Oxford Sci. Pub. 1993.

RELAXED CONSTRAINTS MODEL

A model of relaxed constraints, first proposed by Honneff and Mecking [17] and later developed by Canova, Kocks, and Jonas [18] assumed that after sufficient deformation the grain shape is are no longer equiaxed. Because of their shape (Figure 9.23), the deformation in the center of the grains requires only three slip systems and that five slip systems are required only at the corners. This model becomes progressively more important as the strain increases. Figure 9.24 is a plot of the volume fraction affected as a function of strain for plane strain deformation.

STRAIN HARDENING

To investigate the assumption that strain hardening depends only on $\sum_i |d\gamma_i|$ despite different dislocation interactions in grains of different orientations, plane strain compression experiments were conducted on a large number of aluminum crystals chosen so they would undergo

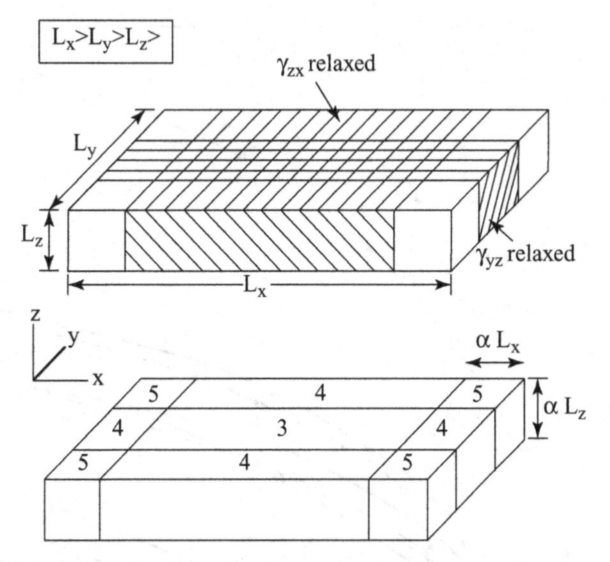

Figure 9.23. Schematic drawing of a distorted grain showing regions in which some constraints are assumed to be relaxed. From W. F. Hosford, *The Mechanics of Crystals and Textured Polycrystals*, Oxford Sci. Pub. 1993.

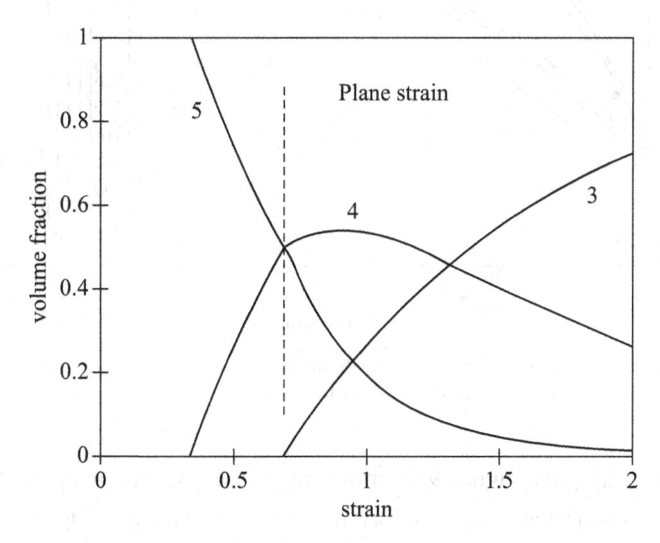

Figure 9.24. How strain affects the volume fraction of material affected by the relaxation model. From W. F. Hosford, *The Mechanics of Crystals and Textured Polycrystals*, Oxford Sci. Pub. 1993.

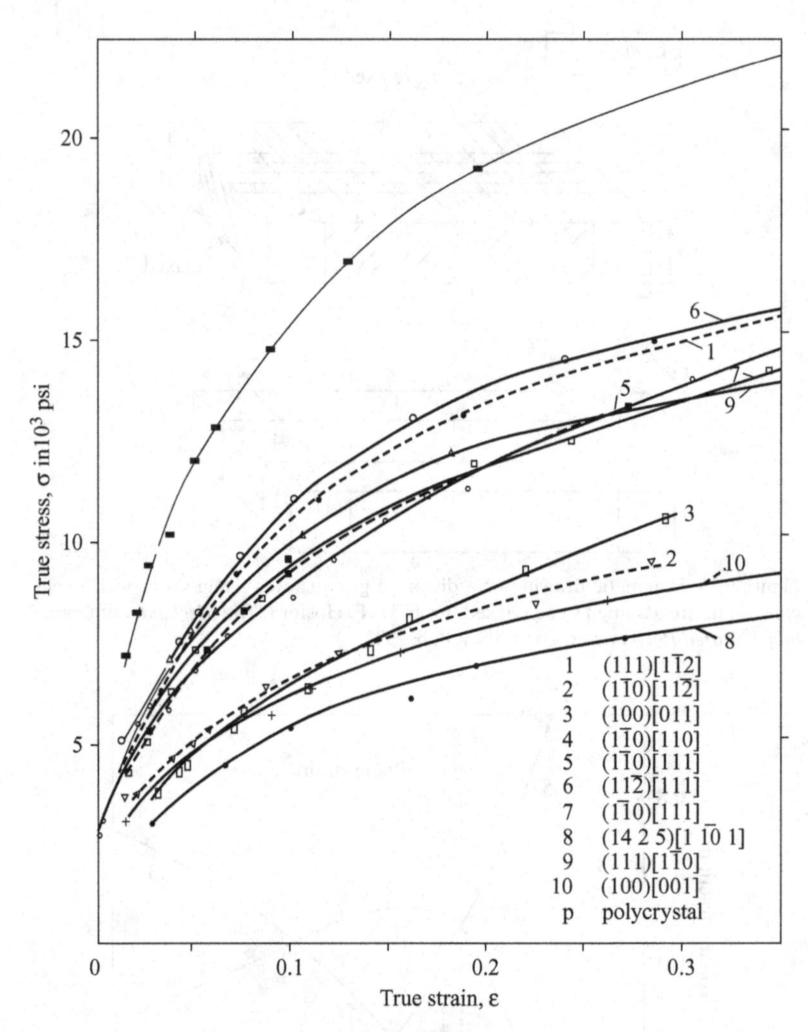

Figure 9.25. True stress strain curves of aluminum single crystals in plane strain compression. From W. F. Hosford, *Acta Met.* v. 14, (1966).

multiple slip [19]. Figure 9.25 shows the resulting σ vs. ε stress strain curves. When these were converted to τ versus γ curves (Figure 9.26), they cluster together, which indicates that the strain hardening is similar for different combinations of multiple slip so the assumption that $\tau = f(\Sigma|\gamma|)$ is valid.

Similar results were found for copper and copper-base alloys.

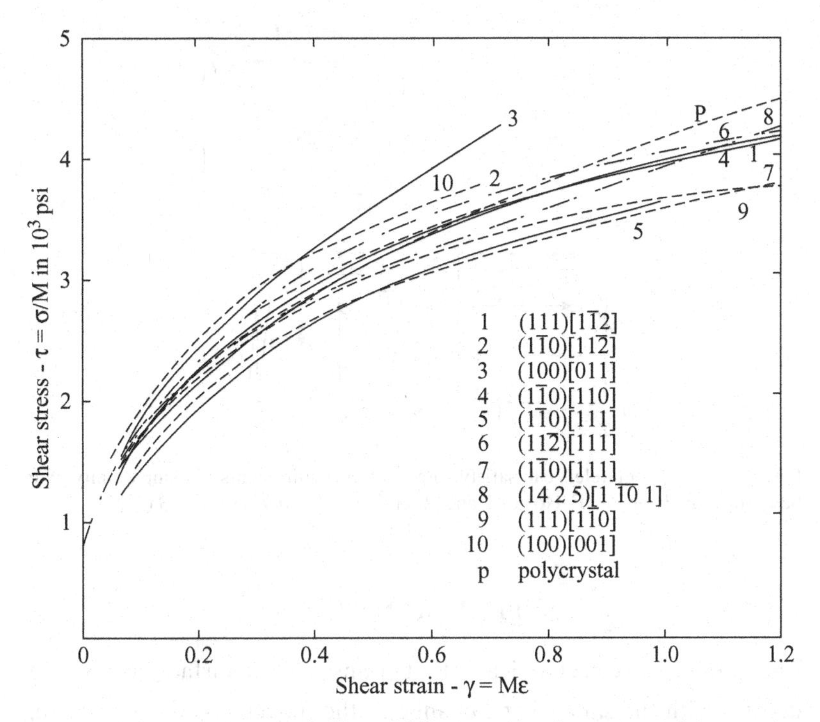

Figure 9.26. Stress strain curves in Figure 9.25 converted to shear stress and shear strain. From W. F. Hosford, *Acta Met.* v. 14, (1966).

SECOND PHASES

With more than one phase, the upper bound for the overall strength is given by

$$\sigma_{av} = f_a\sigma_a + f_b\sigma_b + \cdots$$

where $f_a, f_b, \ldots$ are the volume fractions σ_a and $\sigma_b \ldots$ are the strengths of phases a, b.... It has been suggested [20], however, if one of the phases is in the form of a thin plate and is much stronger than the matrix phase much of the compatibility of the strains, γ_{31} and γ_{12}, can be satisfied by a body rotation as suggested in Figure 9.27.

With $\gamma_{31} = \gamma_{12} = 0$, the effective von Mises effective strain is

$$N = \left[(2/3)\left(d\varepsilon_1^2 + d\varepsilon_2^2 + d\varepsilon_3^2\right) + (1/3)d\gamma_{23}^2\right]^{1/2} \qquad 9.14$$

Similar relaxation can occur for rod-like particles.

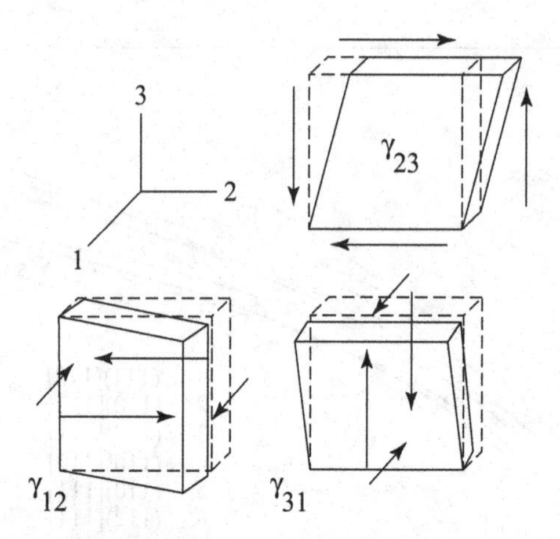

Figure 9.27. Hard platelets can satisfy some of the requirements of compatibility by a body rotation. From W. F. Hosford, and R. H. Zeitsloft, *Met. Trans*, v. 3 (1972).

FREE SURFACE

The orange peal effect indicates that grains at a free surface do not take deform with the same shape change as the polycrystalline aggregate. This leads to a dependence of yield strength on the ratio of grain diameter to specimen diameter [21].

NOTE OF INTEREST

In 1886, Geoffrey Ingram Taylor was born in London. He contributed a steady stream of important papers, principally on fluid mechanics and plasticity. His grandfather was George Boole, who is renown for developing the foundations of what is now called Boolean algebra. Taylor did both his undergraduate studies and graduate research at Cambridge. Taylor never received a doctorate, simply because Cambridge didn't grant them until many years later. During World War I, he did research at Farnsborough on aircraft. After the war, he was appointed lecturer at Cambridge and remained there until his death. Taylor must be regarded as one of the most outstanding scientists of the first half of the 20th century. His work with turbulent flow is fundamental. Taylor's studies of plasticity ranged from testing continuum models, to experiments and analysis of slip in single crystals, to pioneering work that bridged the gap between crystal and continuum mechanics. He is regarded as one of the founders of dislocation theory along with Orowan and Polanyi.

In 1958, four volumes of *The Scientific Papers of Sir Geofrey Ingram Taylor* were published. *Volume I, Mechanics of Solids* contains forty-one papers. The other three volumes contain his writings on fluid mechanics and turbulence.

Though officially retired in 1952, he continued his research. He suffered a stroke in 1972 and died in Cambridge in 1975.

REFERENCES

1. G. Sachs, *Z. Verein Deut. Ing.* v. 72 (1928).
2. H. L Cox and D. J. Sopworth, *Proc. Phys Soc London* v. 49 (1937).
3. E. Schmid and W. Boas, *Kristalllplasticität*, Springer Verlag (1935).
4. G. I. Taylor, *J. Inst. Met.* v. 62 (1938).
5. G. I. Taylor in *Timoshenko Aniv. Vol*, Macmillan (1938).
6. W. F. Hosford, *The Mechanics of Crystals and Textured Polycrystals*, Oxford Sci. *Pub*., 1993.
7. G. Y. Chin and W. L. Mammel, *Trans. TMS-AIME* v. 239 (1967).
8. W. F. Hosford and W. A. Backofen, in Deformation Processing, Proceedings of the 9th Sagamore Army Materials Research Conferenc, Syracuse University Press 1964.
9. J. F. W. Bishop and R. Hill, *Phil Mag., Ser 7* v. 42 (1951).
10. J. F. W. Bishop and R. Hil, *Phil Mag., Ser 7* v. 42 (1951).
11. U. F. Kochs, *Met. Trans* v. 1, (1979).
12. W. F. Hosford, in *Proceedings of the 5th Iinternational Conference on the Strength of Mmetals and Alloys f.* v. 2, Pergamon Press, 1979.
13. R. J. Asaro and A. Needleman, *Acta Met.*, v. 33 (1958).
14. J. F. Peck and D. A. Thomas *TMS-AIME* v 221 (1961).
15. W. F. Hosford, *TMS-AIME*, v. 230 (1964).
16. Van Houtte, in *International Conference on Textures of Materials-7*, Netherlands Society for Materials Science 1984.
17. H. Honneff and H. Mecking, Proc. 5th Int Conf. on Textures of Materials, v. 1, G. Gottstein and K. Lucke, eds. (1978).

18. G. R. Canova, U. F. Kocks and J. J. Jonas, *Acta Met.* v 32 (1984).
19. W. F. Hosford, *Acta Met.* v. 14 (1966).
20. W. F. Hosford, and R. H. Zeitsloft, *Met. Trans* v. 3 (1972).
21. R. L. Fleischer and W. F. Hosford Jr., *Trans. AIME* v. 221 (1961).

10

PENCIL-GLIDE CALCULATIONS OF YIELD LOCI

INTRODUCTION

For pencil glide, the five independent slip variable necessary to produce an arbitrary shape change can be the amount of slip in a given direction and the orientation of the plane (angle of rotation about the direction). There are two possibilities for the five variables: Either three or four active slip directions can be active. Chin and Mammel [1] used a Taylor type analysis for combined slip on {110}, {123}, and {112} planes, finding that M_{av} for axially symmetric flow = 2.748 (Figure 10.1). Hutchinson [2] approximated pencil glide by assuming slip on a large, but finite number of slip planes. Both of these analyses used the least work approach of Taylor. Penning [3] described a least-work solution considering the possibility of both three and four active slip directions. Parniere and Sauzay [4] described a least work solution.

METHOD OF CALCULATION

Piehler et al [5, 7, 8] used a Bishop and Hill-type approach, by considering the stress states capable of activating enough slip systems. Explicit expressions were derived for the stress states in the case of four active slip directions. Instead of explicit solutions for the case of

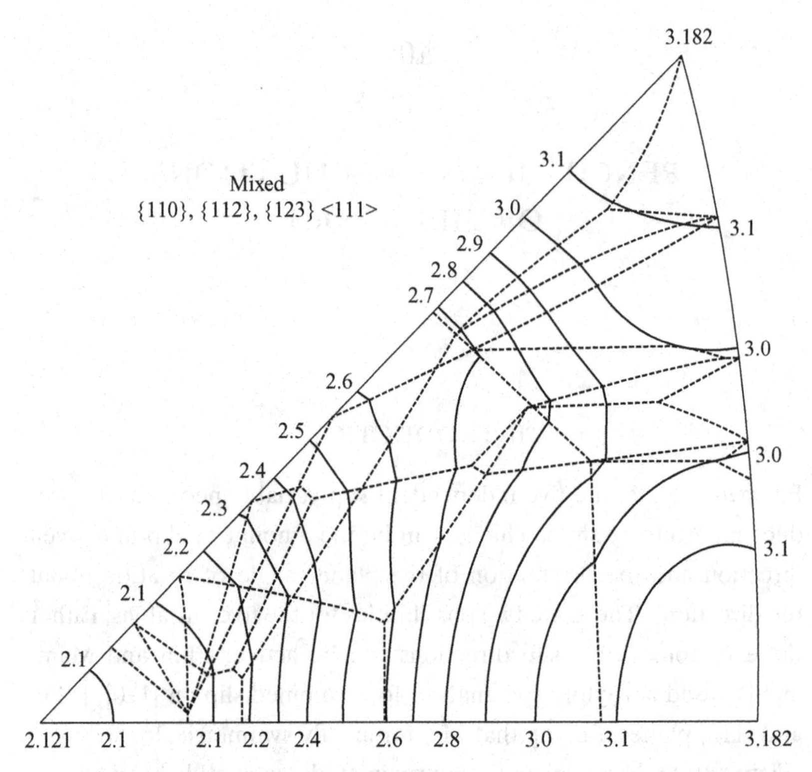

Figure 10.1. The orientation dependence of M for axially symmetric flow with slip on <111>{110}, <111>{112} and <111>{123}. From G. Y. Chin, W. L. Mammel and M. J. Dolan *TMS-AIME*, v 239 (1967).

three active slip directions, a limited number of specific cases were considered. The stress states are:

Expressing the shear strains, γ_a, γ_b, γ_c, and γ_d, on the 1, 2′, and 3′ axes

$$A' = -(3/\sqrt{2})\varepsilon_{2'} = \gamma_b\cos\psi_b + \gamma_d\cos\psi_d \quad 10.1$$

$$B' = (3/\sqrt{2})\varepsilon_{3'} = \gamma_a\cos\psi_a + \gamma_c\cos\psi_c \quad 10.2$$

$$C' = (\sqrt{6/2})\gamma_{2'3'} = -\gamma_a\sin\psi_a + \gamma_b\sin\psi_b - \gamma_c\sin\psi_c + \gamma_a\sin\psi_d \quad 10.3$$

$$D' = 3\gamma_{3'1} = \gamma_a\cos\psi_a + \sqrt{3}\gamma_b\sin\psi_b - \gamma_c\cos\psi_c - \sqrt{3}\gamma_d\sin\psi_d \quad 10.4$$

$$E' = 3\gamma_{12'} = \sqrt{3}\gamma_a\sin\psi_b + \gamma_b\cos\psi_b + \sqrt{3}\gamma_c\sin\psi_c - \gamma_d\cos\psi_d, \quad 10.5$$

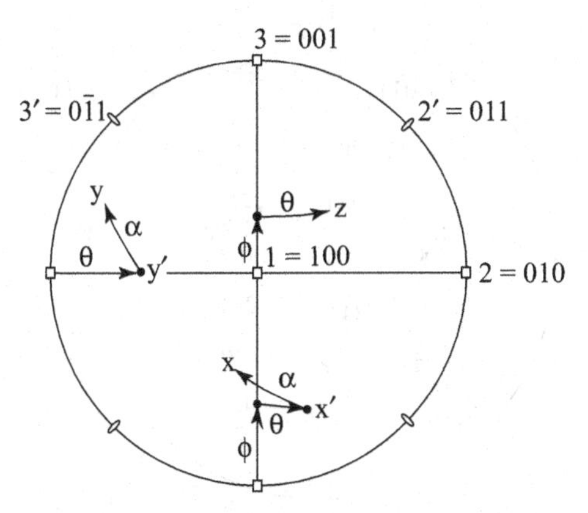

Figure 10.2. Stereographic representation of the axes. From R. W. Logan and W. F. Hosford, *Int. J. Mech. Sci.*, v. 22, 1980.

where ψ_a, ψ_b, ψ_c, and ψ_d describe the rotations of the slip planes about the <111> directions.

Logan and Hosford [8] combined the maximum virtual work and the least shear analyses. Piehler's stress states were used for the cases in which four of the slip directions are active and Taylor's least shear approach for the cases of three active slip directions. To determine which solution was appropriate (three or four slip directions), the three slip direction solution was first tried and checked to see if it were appropriate. If it is not, the four direction solution was found.

First a set of external strains, ε_x, ε_y, ε_z, γ_{yz}, γ_{zx}, γ_{xy}, was assumed and the crystal orientation described by angles ϕ, θ and α relative to the cubic crystal axes, 1, 2, and 3 as shown in Figure 10.2. The strains on the x, y, z axes are transformed to the 1, 2, 3 axes and finally to the 1, 2′, 3′ axis system where 2′ = [011] and 3′ = [01$\bar{1}$] as shown in Figure 10.3.

With three active slip directions, there are six unknowns on the right-hand side of equations 10.1 though 10.5. Letting one of the shear strains be zero, (that is, γ_a), the equations simplify to

$$\gamma_c = B'/\cos\psi_c \qquad 10.6$$

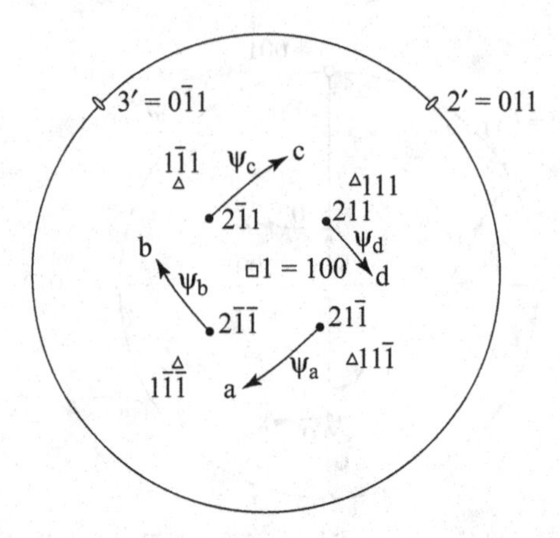

Figure 10.3. Stereographic representation of the slip-plane normals, a, b, c and d and the angular rotation of these about the slip directions. *Int. J. Mech. Sci.*, v. 22, 1980.

$$\gamma_b = \sqrt{[(D_p + C_p)^2 + (E_p + A')^2]/2} \qquad 10.7$$

$$\gamma_d = \sqrt{[(D_p - C_p)^2 + (E_p - A')^2]/2} \qquad 10.8$$

where $C_p = C' + \gamma_c \sin\psi_c$, $D_p = (D' + B')/\sqrt{3}$, and $E_p = E' - \sqrt{3}\gamma_c \sin\psi_c$. Assuming a value for γ_c, equations 10.6, 10.7, and 10.8 may be solved for γ_b, γ_c, and γ_d, and $\gamma_T = |\gamma_b| + |\gamma_c| + |\gamma_d|$. This process is repeated, varying γ_c, to find lowest value of γ_T.

Then the corresponding stress states, $A = \sigma_2 - \sigma_3$, $B = \sigma_3 - \sigma_1$, $C = \sigma_1 - \sigma_2$, $F = \tau_{23}$, $G = \tau_{31}$, $H = \tau$, can be found from

$$\tau_a \sin\psi_a = (A - G - H)/\sqrt{6} \qquad 10.9$$

$$\tau_a \cos\psi_a = (C - B - H + G + 2F)/(3/\sqrt{2}) \qquad 10.10$$

$$\tau_b \sin\psi_b = (-A + G - H)/\sqrt{6} \qquad 10.11$$

$$\tau_b \cos\psi_b = (C - B + H + G - 2F)/(3/\sqrt{2}) \qquad 10.12$$

$$\tau_c \sin\psi_c = (A + G + H)/\sqrt{6} \qquad 10.13$$

$$\tau_c \cos\psi_c = (C - B + H - G + 2F)/(3/\sqrt{2}) \qquad 10.14$$

$$\tau_d \sin\psi_d = (-A - G + H)/\sqrt{6} \qquad 10.15$$

$$\tau_d \cos\psi_d = (C - B - H - G - 2F)/(3/\sqrt{2}), \qquad 10.16$$

which were derived from expressing the four values of τ_i in terms of the stress state and setting $d\tau/d\psi_i = 0$. Now let $\tau_a = \tau_b = \tau_c = \tau_d = 1$ and solve for *A, B, C, D, F, G*, and H. With these, the value of τ_a can be found from equation 10.9. If $|\tau_a| \leq 1$, the solution is appropriate. Otherwise, the remaining cases of three systems must be examined. If none are appropriate, four systems must be active.

Piehler [6] showed that for four systems to operate,

$$F^2(B - C) = G^2(C - A) = H^2(A - B) = 0 \qquad 10.17$$

There are four possible stress states for four active slip directions:

$$\begin{aligned}
1 \quad & A/k = \sqrt{6}(d\varepsilon_1 + d\varepsilon_2)/[2(d\varepsilon_1^2 + d\varepsilon_1 d\varepsilon_2 + d\varepsilon_3^2)] \\
& B/k = -\sqrt{6}(d\varepsilon_1 + d\varepsilon_2)/[2(d\varepsilon_1^2 + d\varepsilon_2^2 + d\varepsilon_3^2)] \\
& C/k = -\sqrt{6}(d\varepsilon_1 - d\varepsilon_2)/[2(d\varepsilon_1^2 + d\varepsilon_1 d\varepsilon_2 + d\varepsilon_3^2)]
\end{aligned}$$

$$\begin{aligned}
\text{IIIa} \quad & A/k = -2\sqrt{6}(d\varepsilon_2 - d\varepsilon_3)/[2(12d\varepsilon_{23}^2 + (d\varepsilon_2 + d\varepsilon_3)^2] \\
& B/k = -2\sqrt{6}(d\varepsilon_3 - d\varepsilon_2)/[2(12d\varepsilon_{23}^2 + (d\varepsilon_2 - d\varepsilon_3)^2] \\
& C/k = 2\sqrt{6}(d\varepsilon_3 - d\varepsilon_2)/[2(12d\varepsilon_{23}^2 + (d\varepsilon_2 - d\varepsilon_3)^2] \\
& F/k = 6\sqrt{6}(d\varepsilon_{23})/[2(12d\varepsilon_{23}^2 + (d\varepsilon_2 - d\varepsilon_3)^2]
\end{aligned}$$

$$\begin{aligned}
\text{IIIb} \quad & A/k = \sqrt{6}(d\varepsilon_1 - d\varepsilon_3)/[2(12d\varepsilon_{31}^2 + (d\varepsilon_3 - d\varepsilon_1)^2] \\
& B/k = \sqrt{6}(d\varepsilon_3 - d\varepsilon_1)/[2(12d\varepsilon_{31}^2 + (d\varepsilon_3 - d\varepsilon_1)^2] \\
& C/k = \sqrt{6}(d\varepsilon_1 - d\varepsilon_3)/[2(12d\varepsilon_{31}^2 + (d\varepsilon_3 - d\varepsilon_1)^2] \\
& G/k = 6\sqrt{6}d\varepsilon_{31}/[2(12d\varepsilon_{31}^2 + (d\varepsilon_3 - d\varepsilon_1)^2]
\end{aligned}$$

and

$$\begin{aligned}
\text{IIIc} \quad & A/k = \sqrt{6}(d\varepsilon_2 - d\varepsilon_1)/[2(12d\varepsilon_{12}^2 + (d\varepsilon_1 - d\varepsilon_{12})^2] \\
& B/k = \sqrt{6}(d\varepsilon_2 - d\varepsilon_1)/[2(12d\varepsilon_{12}^2 + (d\varepsilon_1 - d\varepsilon_{12})^2] \\
& C/k = \sqrt{6}(d\varepsilon_1 - d\varepsilon_2)/[2(12d\varepsilon_{12}^2 + (d\varepsilon_1 - d\varepsilon_{12})^2] \\
& H/k = 6\sqrt{6}d\varepsilon_{12}/[2(12d\varepsilon_{12}^2 + (d\varepsilon_1 - d\varepsilon_{12})^2]
\end{aligned}$$

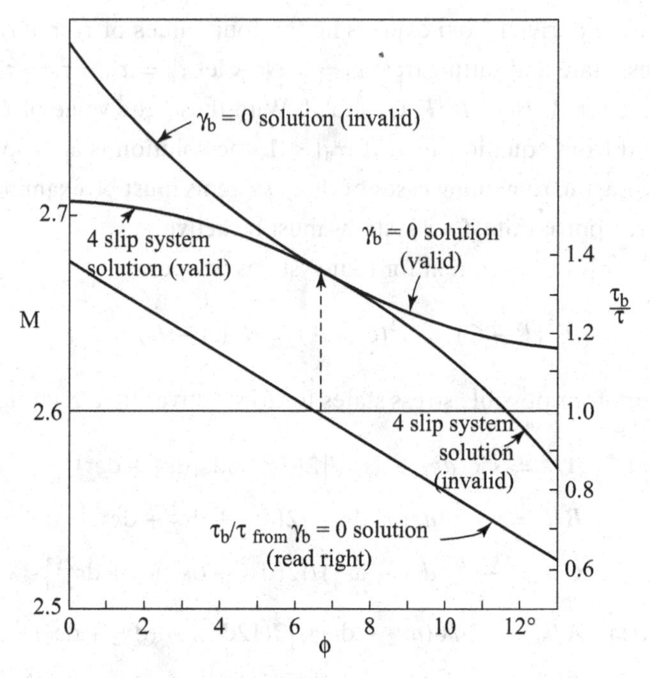

Figure 10.4. Calculated Taylor factors for the three system, $\gamma_b = 0$, solution, and the four system solution. The $\gamma_b = 0$ solution is valid for $\tau_b/\tau \leq 1$ and the four system solution is valid for $\tau_b/\tau \geq 1$. From *Int. J. Mech. Sci.*, v. 22, 1980.

The appropriate stress state is that one which gives the largest value of dw, where

$$\mathrm{d}w = -B\mathrm{d}\varepsilon_1 + A\mathrm{d}\varepsilon_2 + F\mathrm{d}\gamma_{23} + G\mathrm{d}\gamma_{31} + H\mathrm{d}\gamma_{12} \qquad 10.18$$

This maximum virtual work solution was checked for consistency by solving equations 10.1 through 10.5 for the shear strains. If it is correct, the sum of the absolute magnitudes will equal the value of dw in equation 10.18 divided by τ.

The value of dw for the four system solution is always lower than the solution for the three system solution as shown in Figure 10.4.

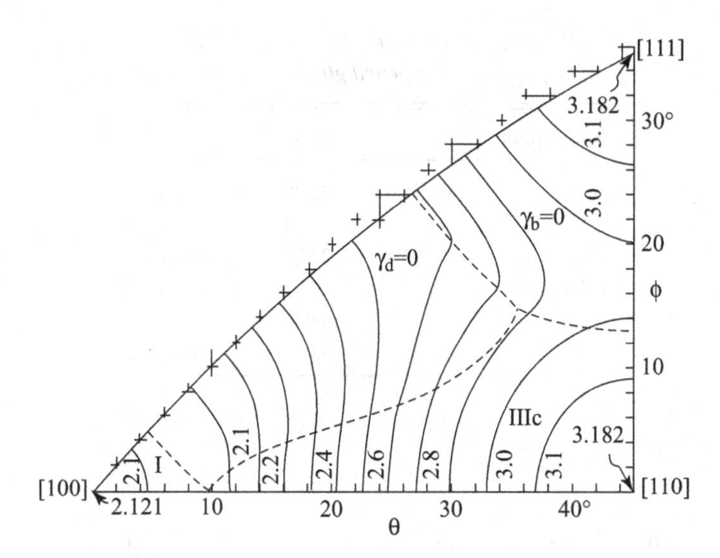

Figure 10.5. Orientation dependence of the Taylor factor, *M* for axially symmetric deformation. From *Int. J. Mech. Sci.*, v. 22, 1980.

CALCULATION RESULTS

For axially symmetric deformation, the basic orientation triangle is divided into four regions as shown in Figure 10.5. The three-system solution is appropriate in two of the regions and the four system solution in the other two. The Taylor factor averaged over all orientations is 2.7398 ± 0.0016. This is somewhat higher than Piehler's value of 2.748.

Table 10.1 gives the calculated values of σ_x/τ and σ_y/τ for pencil glide and Figure 10.6 shows these data in a Lode variable plot.

Figures 10.7, 10.8, and 10.9 show the yield loci calculated for textures with <100>, <110>, and <111> sheet normals and rotational symmetry about those normals approximated by orientations rotated 3 degrees.

The results of upper-bound calculations the ratios X, λ, ψ, ξ and β, are plotted in Figures 10.10 through 10.14 as a function of the calculated R-value.

Table 10.1. *Calculated values of σ_x/τ and σ_y/τ for pencil glide*

Strain path $\rho = \varepsilon_y/\varepsilon_z$	σ_x/τ	σ_y/τ
0.500	2.740	0.000
0.4000	2.834	0.2283
0.300	2.928	0.468
0.200	2.996	0.728
0.100	3.040	1.115
0.000	3.0647	1.539

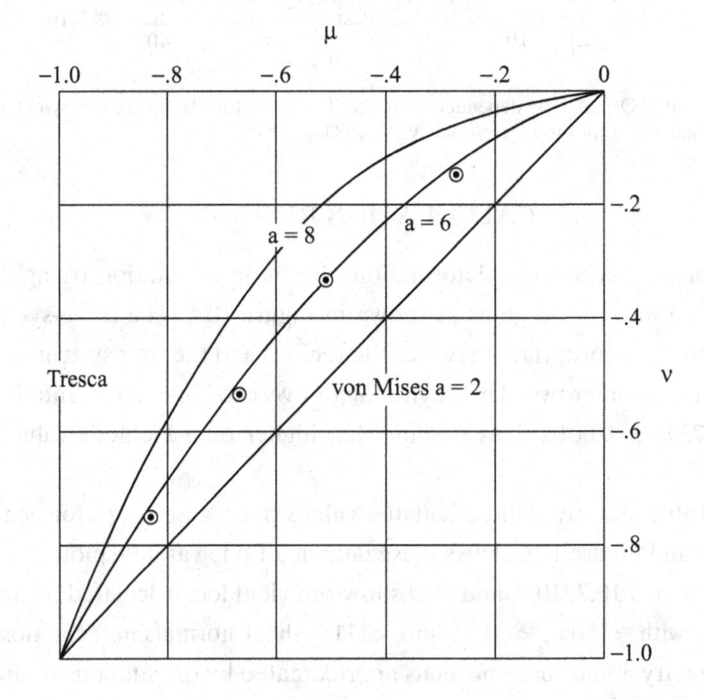

Figure 10.6. Lode variable plot of calculated stress and strain together with the predictions of equations 4.10 and 4.15 with a = 6 and 8. The solid points are the calculations in Table 10.1.

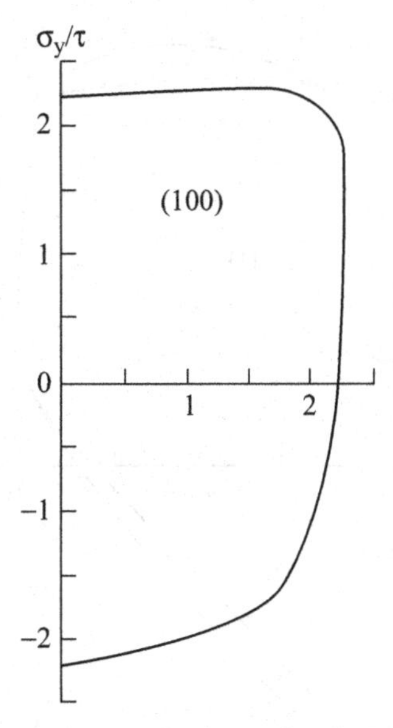

Figure 10.7. Yield locus for texture that is rotationally symmetric about <100>.

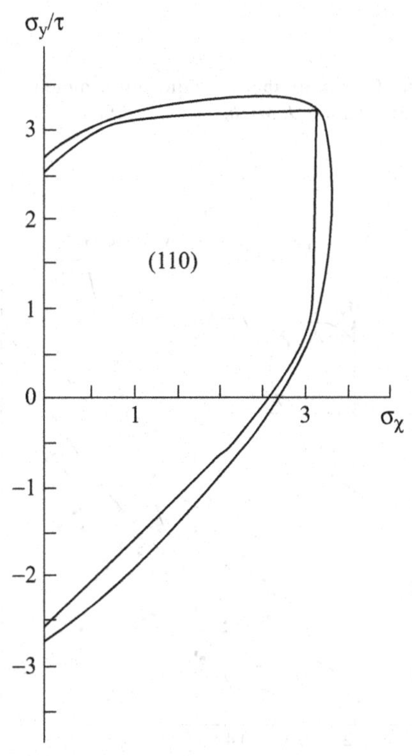

Figure 10.8. Yield locus for texture that is rotationally symmetric about <110>. Inner locus is from Piehler [5] From From R. W. Logan and W. F. Hosford, *Int. J. Mech. Sci.*, v. 22, 1980.

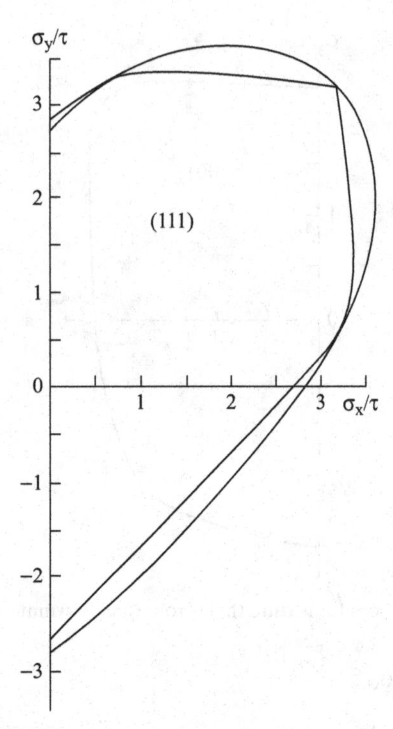

Figure 10.9. Yield locus for texture that is rotationally symmetric about <111>. Inner locus is from Piehler [5] From *Int. J. Mech. Sci.*, v. 22, 1980.

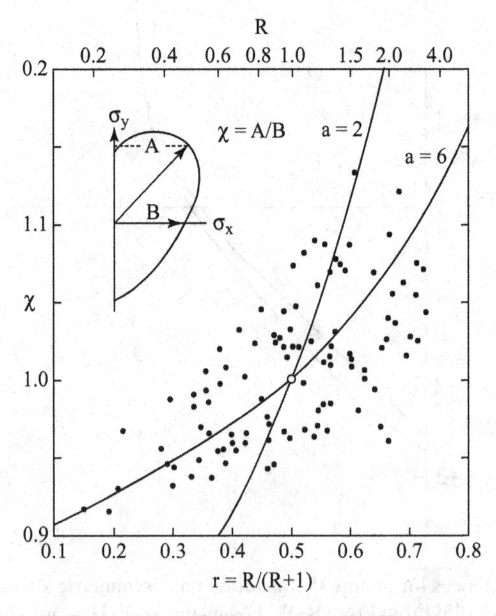

Figure 10.10. Dependence of the strength ratio, X on the strain ratio.

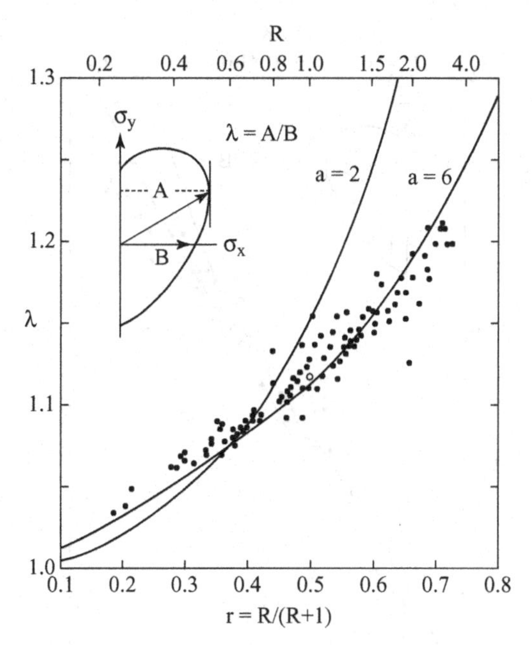

Figure 10.11. Dependence of the strength ratio, λ on the strain ratio, *R*. From R. W. Logan and W. F. Hosford, *Int. J. Mech. Sci.*, v. 22, 1980.

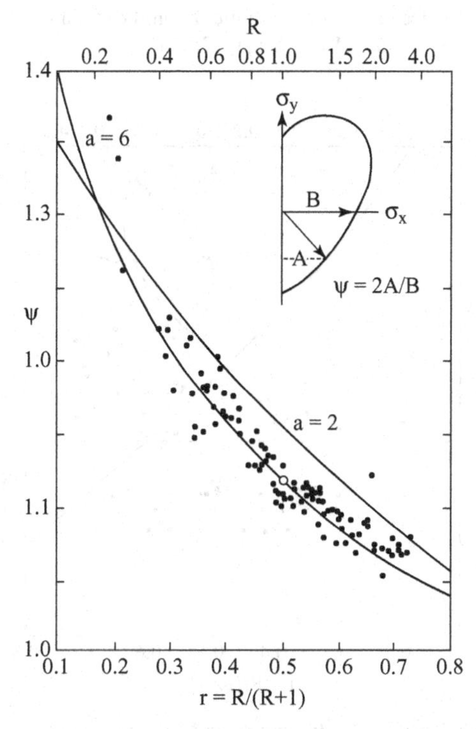

Figure 10.12. Dependence of the strength ratio, ψ, on the strain ratio, *R*. From R. W. Logan and W. F. Hosford, *Int. J. Mech. Sci.*, v. 22, 1980.

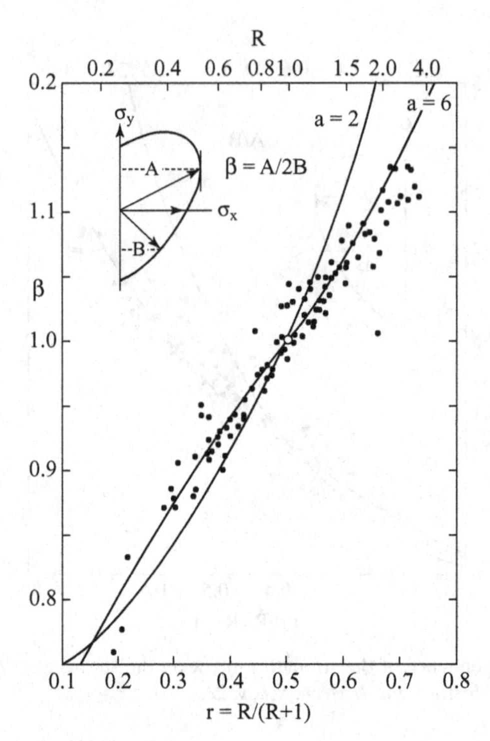

Figure 10.13. Dependence of the strength ratio, β, on the strain ratio, R. From R. W. Logan and W. F. Hosford, *Int. J. Mech. Sci.*, v. 22, 1980.

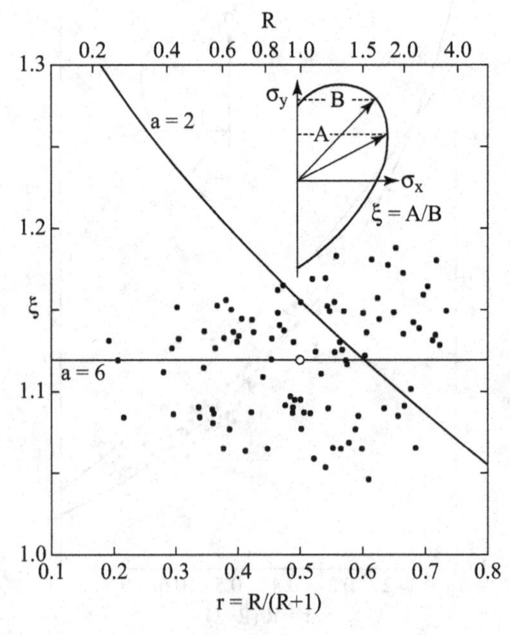

Figure 10.14. Dependence of the strength ratio, ξ, on the strain ratio, R. From R. W. Logan and W. F. Hosford, *Int. J. Mech. Sci.*, v. 22, 1980.

REFERENCES

1. G. Y. Chin, W. L. Mammel, and M. J. Dolan *TMS-AIME* v. 239 (1967).
2. J. W. Hutchinson, *J Mech. Phys. Solids* v. 12 (1964).
3. P. Penning, *Met Trans* v. 7A (1976).
4. P. Parnier and C. Sauzay, *Mat. Sci. Engr* v. 22 (1976).
5. H. R. Piehler and W. A. Backofen, in *Textures in Research and Practice*, J. Grewen and G. Wasserman eds., Springer Verlag (1961).
6. H. R. Piehler and W. A. Backofen, *Met. Trans.* v. 2 (1971).
7. J. M. Rosenberg and H. Piehler, *Met. Trans.* v. 2 (1971).
8. R. W. Logan and W. F. Hosford, "Upper-Bound Anisotropic Yield Locus Calculations Assuming <111>-Pencil Glide," *Int. J. Mech. Sci.* v. 22 (1980).

11

MECHANICAL TWINNING AND MARTENSITIC SHEAR

TWINNING

Many crystalline materials can deform by twinning as well as by slip. Mechanical twinning, like slip, occurs by shear. A twin is a region of a crystal in which the orientation of the lattice is a mirror image of that in the rest of the crystal. Twins may form during recrystallization (*annealing twins*), but the concern here is formation of twins by uniform shearing (*mechanical twinning*) as illustrated in Figure 11.1. In this figure, plane 1 undergoes a shear displacement relative to plane 0 (the mirror plane). Then, plane 2 undergoes the same shear relative to plane 1, and plane 3 relative to plane 2, and so on. The net effect of the shear between each successive plane is to reproduce the lattice, but with the new (mirror image) orientation.

Both slip and twinning are deformation mechanisms that involve shear displacements on specific crystallographic planes and in specific crystallographic directions. However, there are major differences.

1. With slip, the magnitude of the shear displacement on a plane is variable, but it is always an integral number of interatomic repeat distances, nb where b is the Burgers vector. Slip occurs on only a few of the parallel planes separated by relatively large distances. With twinning, on the other hand, the shear displacement is a fraction of an inter-atomic repeat distance and every atomic plane shears relative to its neighboring plane.

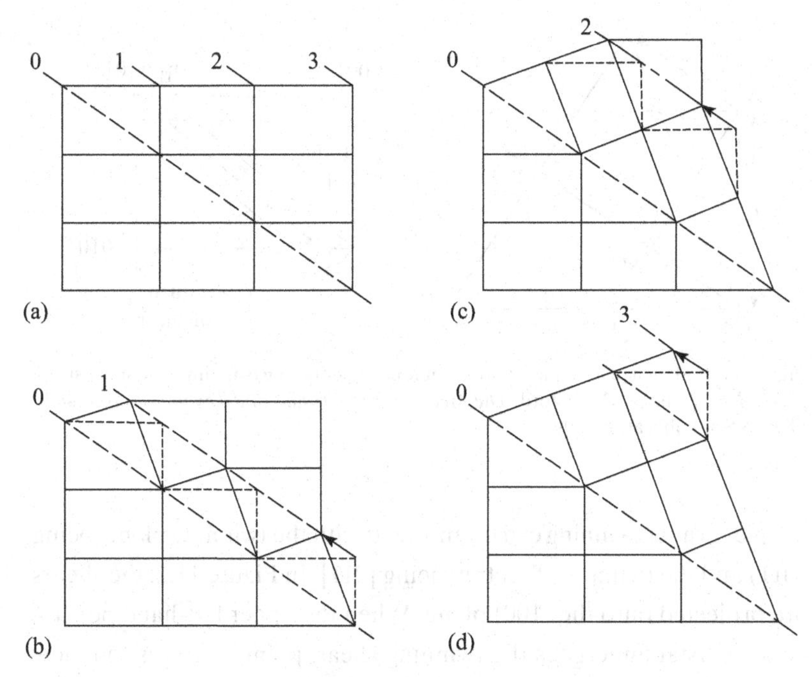

Figure 11.1. Formation of twins by shearing between each parallel plane of atoms. From W. F. Hosford, *The Mechanics of Crystals and Textured Polycrystals,* Oxford Sci. Publication (1993).

2. The twinning shear is always directional in the sense that if shear in one direction produces a twin, shear in the opposite direction will not. Twinning in fcc crystals occurs by shear on the (111) plane in the $[11\bar{2}]$ direction but not by shear in the $[\bar{1}\bar{1}2]$ direction. In fcc metals, slip can occur on a (111) plane in either the $[\bar{1}10]$ or the $[1\bar{1}0]$ directions.
3. With slip the lattice rotation is gradual. Twinning causes an abrupt reorientation.

LATTICE SHEARS

Many twins can be regarded as forming by a shearing that creates a mirror image in an orthorhombic cell as shown in Figure 11.2. This

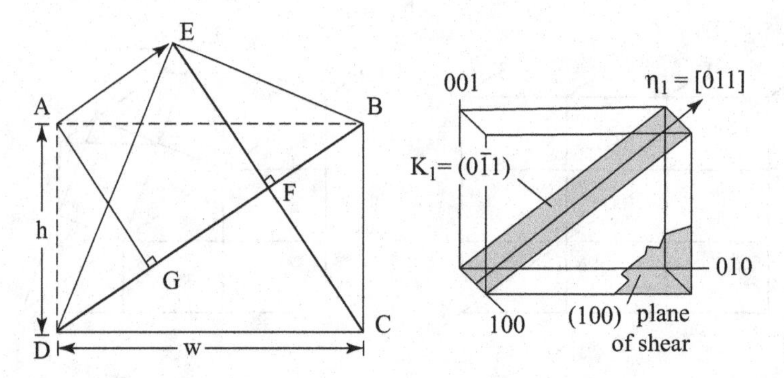

Figure 11.2. Sketch of a simple orthorhombic cell undergoing twinning. The shear strain is $\bar{A}\bar{E}/\bar{E}\bar{F}$. From W. F. Hosford, *The Mechanics of Crystals and Textured Polycrystals,* Oxford Sci. Publication (1993).

simple form of twinning can be analyzed with the twinning plane, being $(01\bar{1})$ and the twinning direction being [011]. In Figure 11.2, the shears are projected onto the (100) plane. When the upper left-hand portion of the crystal undergoes the twinning shear, point A moves to a new position, E. The shear strain, γ, is defined as $\bar{A}\bar{E}/\bar{E}\bar{F}$. The length, $\bar{E}\bar{F}$, can be related to w and to h by noting that the triangle EFB is similar to triangle DCB, so that $\bar{E}\bar{F}/\bar{D}\bar{C} = \bar{E}\bar{B}/\bar{D}\bar{B}$. Substituting $w = \bar{D}\bar{C}$, $h = \bar{E}\bar{B} = \bar{B}\bar{C}$, and $\bar{D}\bar{B} = \sqrt{}(w^2+ h^2)$, $\bar{E}\bar{F} = wh/\sqrt{}(w^2+ h^2)$. Because triangle EFB is similar to triangle DCB, $\bar{D}\bar{C}/\bar{E}\bar{B} = \bar{B}\bar{C}/\bar{D}\bar{C}$, so the length $\bar{D}\bar{C} = h^2/\sqrt{}(w^2+ h^2)$. Therefore, $\bar{A}\bar{E} = \bar{D}\bar{B} - 2\,\bar{F}\bar{B} = \sqrt{}(w^2 + h^2) - 2h^2/\sqrt{}(w^2 + h^2)$. Substituting, $\gamma = \bar{A}\bar{E}/\bar{E}\bar{F} = [\sqrt{}(w^2 + h^2) - 2(h^2/\sqrt{}(w^2 + w^2)\}/[wh/\sqrt{}(w^2 + h^2)]$. Simplifying,

$$\gamma = w/h - h/w. \qquad 11.1$$

If the center of the cell (point H in Figure 11.3) is connected to points A and E as shown in Figure 11.3, it is clear that line $\bar{H}\bar{A}$ in the parent is equivalent to line $\bar{H}\bar{E}$ in the twin. The angle of reorientation, θ, of the [010] direction and (001) plane is given by

$$\tan(\theta/2) = h/w. \qquad 11.2$$

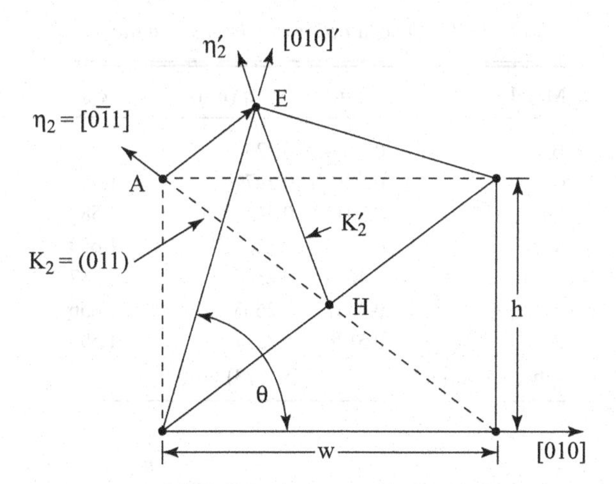

Figure 11.3. Determination of the reorientation angle, θ, of the [010] direction and the twinning elements, K_1 and K_2. From W. F. Hosford, *The Mechanics of Crystals and Textured Polycrystals,* Oxford Sci. Publication (1993).

Table 11.1 gives the values of h and w in equations 11.1 and 11.2 for several crystal structures.

TWINNING IN fcc METALS

Twinning in fcc metals occurs on {111} planes in <11$\bar{2}$> directions. The atomic movements are shown in Figures 11.4 and 11.5. The shear strain can be calculated from equation 11.1 by substituting $w = a\sqrt{2}$ and $h = a$, $\gamma = a\sqrt{2}/a - a/(a\sqrt{2}) = \sqrt{2}/2 = 0.707$. From equation 11.2, the angle of reorientation of the [001] direction is $\theta = 2\arctan(h/w) = 2\arctan(1/\sqrt{2}) = 70.5°$.

Table 11.1. *Values of* h, w *and* γ

Crystal structure	h	w	γ
fcc and dia. cubic	a	$a\sqrt{2}$	$\sqrt{2}/2$
Bcc	a	$a/\sqrt{2}$	$\sqrt{2}/2$
Hcp	c	$a\sqrt{3}$	$c/(\sqrt{3}a) - \sqrt{3}a/c$

Table 11.2. *The* c/a *ratios of various hcp metals*

Metal	c (nm)	a (nm)	c/a
Be	0.3584	0.2286	1.568
Cd	0.5617	0.2979	1.886
Hf	0.5042	0.3188	1.582
Mg	0.5210	0.3209	1.624
Ti	0.4683	0.2590	1.587
Zn	0.4947	0.2665	1.856
Zr	0.5148	0.3231	1.593
spherical atoms		$\surd(8/3) = 1.633$	

TWINNING IN bcc METALS

Figure 11.6 is a plan view of the (110) plane in a bcc metal. The upper right half has undergone the twinning shear on the $(\bar{1}12)[1\bar{1}1]$ system. The shear strain is $\gamma = \surd 2/2$ and produces a tensile elongation parallel to [001]. Thus, the deformation is equal in magnitude to that

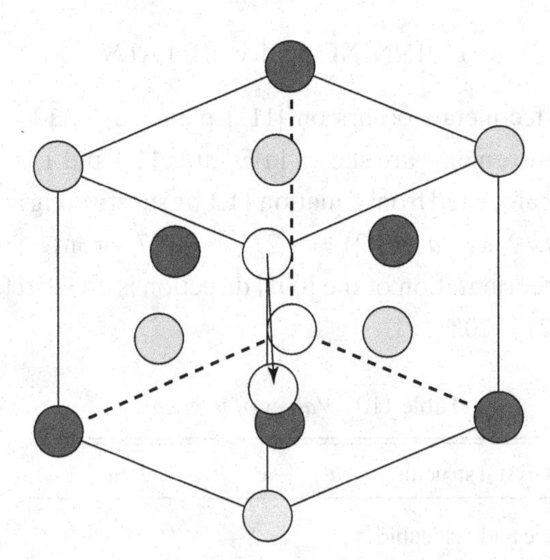

Figure 11.4. Atomic displacement in $\{111\}\langle 11\bar{2}\rangle$ twinning of fcc metals. From W. F. Hosford, *The Mechanics of Crystals and Textured Polycrystals,* Oxford Sci. Publication (1993).

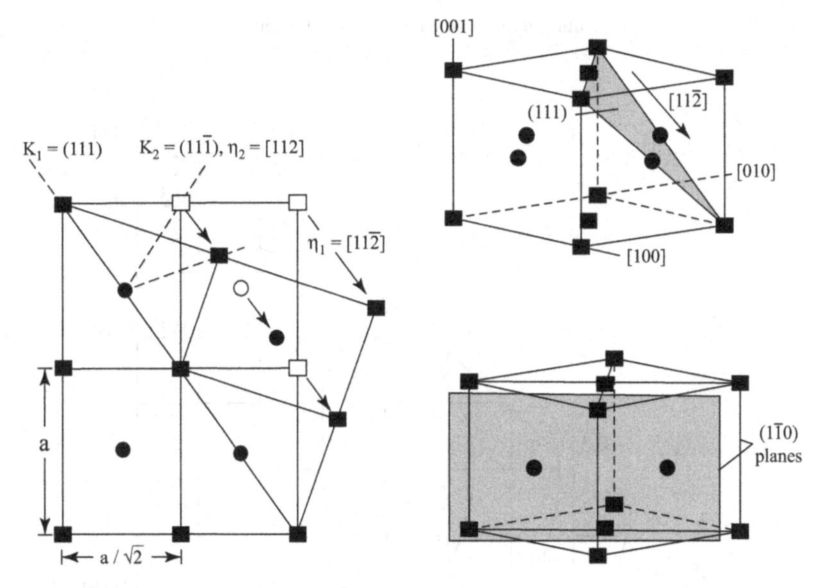

Figure 11.5. Plan view of $\{111\}\langle 11\bar{2}\rangle$ twinning in fcc crystals. The atom positions are projected onto the shear plane, $(1\bar{1}0)$. From W. F. Hosford, *The Mechanics of Crystals and Textured Polycrystals,* Oxford Sci. Publication (1993).

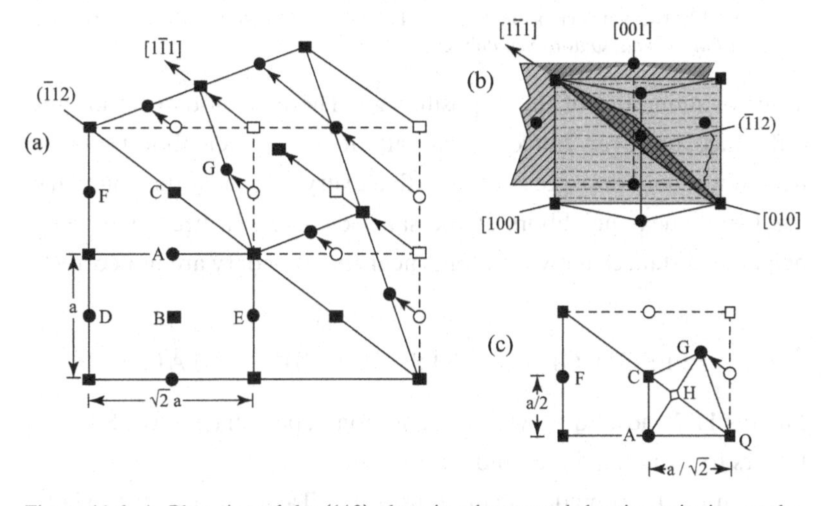

Figure 11.6. A. Plan view of the (110) plane in a bcc crystal showing twinning on the $(\bar{1}12)[1\bar{1}1]$ system in the upper right half of the figure. The atoms in several unit cells are indicated. Those in one (110) plane are shown as squares while those one plane forward and one plane back are indicated by circles. B. Perspective sketch showing the twinning elements. C. Nearest neighbor distances. The distance between *A* and *G* is too small. From W. F. Hosford, *The Mechanics of Crystals and Textured Polycrystals,* Oxford Sci. Publication (1993).

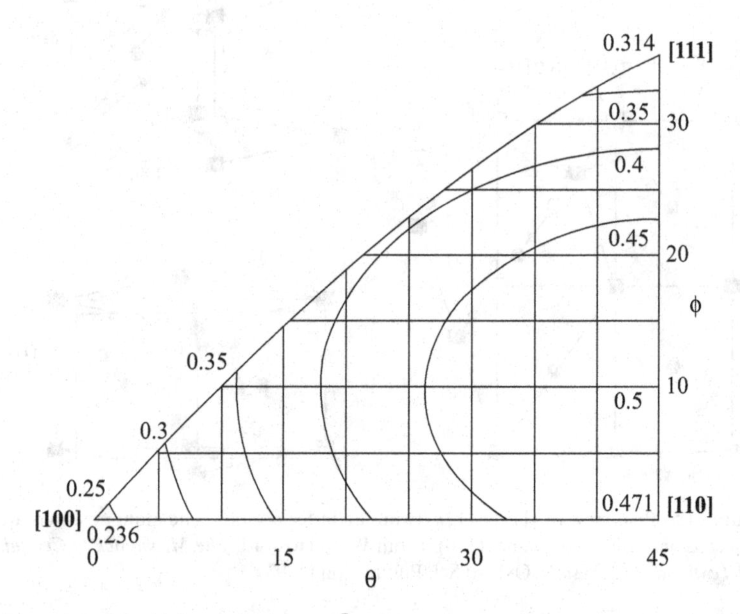

Figure 11.7. Schmid factor for bcc <11$\bar{1}$>{112} twinning in uniaxial compression and fcc {$\bar{1}\bar{1}$1}<112> twinning in tension. From W. F. Hosford, *The Mechanics of Crystals and Textured Polycrystals,* Oxford Sci. Publication (1993).

in fcc twinning but of the opposite sign. There is another important difference, however. In fcc crystals, all the nearest neighbor distances near a twin boundary are correct so that every atom near the boundary has twelve near neighbors at the same distance. In bcc some near-neighbor distances between atoms near the boundary are not correct.

EFFECTS OF TWINNING IN CUBIC METALS

Figures 11.7 and 11.8 show the orientation dependence of the Schmid factors for twinning in fcc and bcc metals.

Chin et al. [2] made calculations of the Taylor factors for axially symmetric flow by twinning under uniaxial tension and compression (Figures 11.9 and 11.10).

Figure 11.11 shows the asymmetric yield loci calculated for randomly oriented fcc and bcc polycrystals that deform only by twinning.

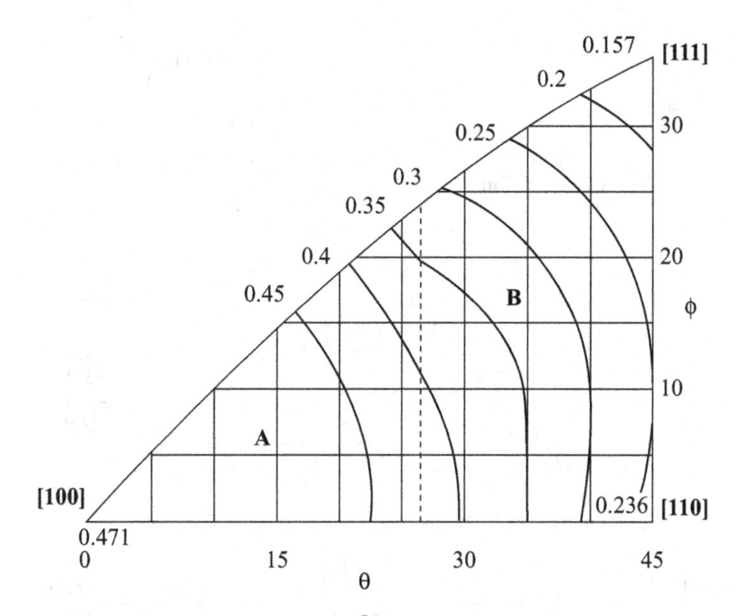

Figure 11.8. Schmid factor for bcc <11$\bar{1}$>{112} twinning in uniaxial tension and fcc {$\bar{1}\bar{1}$1}<112> twinning in compression. From W. F. Hosford, *The Mechanics of Crystals and Textured Polycrystals,* Oxford Sci. Publication (1993).

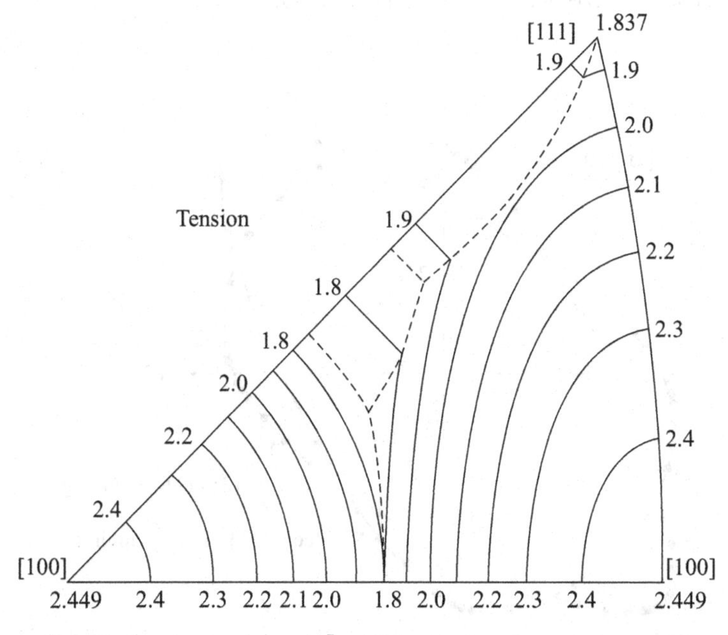

Figure 11.9. Taylor factor bcc <11$\bar{1}$>{112} twinning in uniaxial tension and fcc {$\bar{1}\bar{1}$1}<112> twinning in compression. From W. F. Hosford, *The Mechanics of Crystals and Textured Polycrystals,* Oxford Sci. Publication (1993).

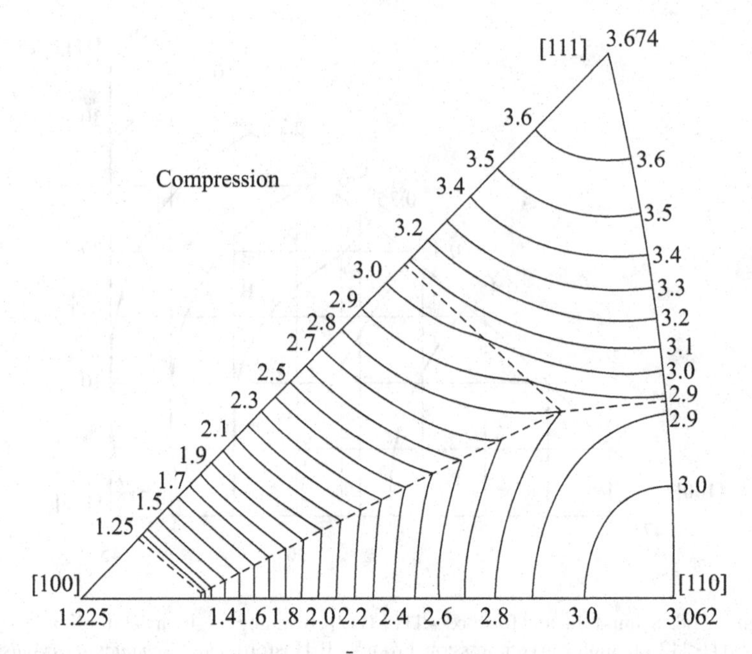

Figure 11.10. Taylor factor bcc <11$\bar{1}$>{112} twinning in uniaxial tension and fcc {$\bar{1}\bar{1}$1}<112> twinning in compression. From W. F. Hosford, *The Mechanics of Crystals and Textured Polycrystals,* Oxford Sci. Publication (1993).

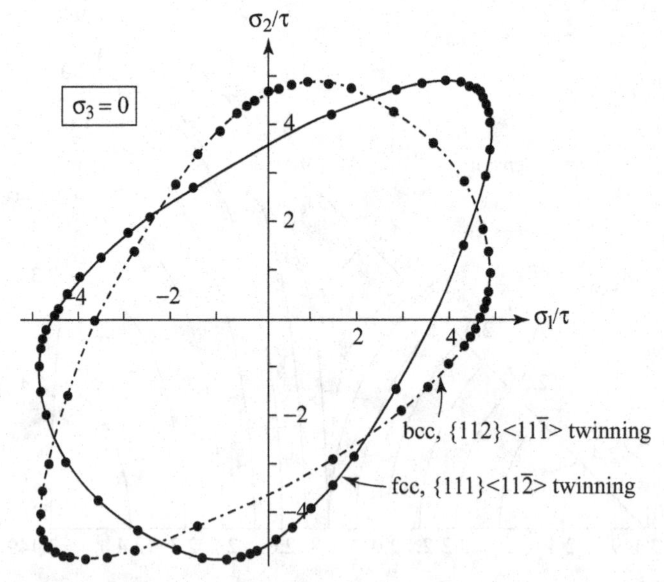

Figure 11.11. Calculated yield loci for randomly oriented fcc and bcc polycrystals deforming only by twinning. From W. F. Hosford and T. J. Allen, *Met Trans,* v. 4 (1973).

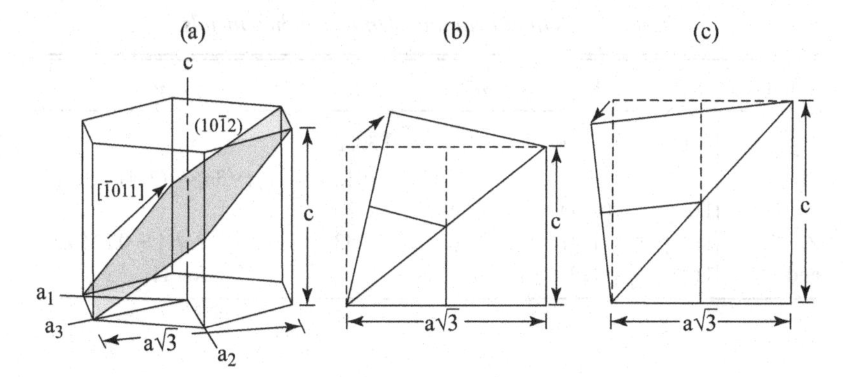

Figure 11.12. A. Perspective sketch showing {10$\bar{1}$2}<10$\bar{1}$1> twinning in hcp crystals. B. For $c/a < \sqrt{3}$, the direction of shear is [$\bar{1}$011] and twinning will occur under tension parallel to the c-axis. C. For $c/a > \sqrt{3}$, the direction of shear is [10$\bar{1}\bar{1}$] and twinning will occur under compression parallel to the c-axis. From W. F. Hosford, *The Mechanics of Crystals and Textured Polycrystals,* Oxford Sci. Publication (1993).

TWINNING OF hcp METALS

Because of limited slip systems, twinning is much more important in the hcp metals than in the cubic metals. The number of slip systems in the hcp metals is limited. All the easy slip <11$\bar{2}$0> directions are perpendicular to the c-axis and therefore slip doesn't produce any elongation or shortening parallel to the c-axis.

The most common twinning system in hcp metals is {10$\bar{1}$2}< $\bar{1}$011>. This is a simple mode of the form in Figure 11.12 and equations 11.1 and 11.2. Figure 11.12 shows that the direction of shear associated with this twinning system depends on the c/a ratio. These are given in Table 11.3 for a number of hcp metals. Only zinc and cadmium have *c/a* ratios greater than $\sqrt{3}$. Therefore, only these metals twin when compression is applied parallel to the c-axis (or tensile stresses perpendicular to it). For all the other hcp metals $c/a < \sqrt{3}$, so twins are formed by tensile stresses parallel to the c-axis (or compressive stresses perpendicular to it). It has been found that the frequency of twins in Cd-Mg solid solution alloys decreases as the composition approaches that for which $c/a = \sqrt{3}$.

Table 11.3. *Other twinning elements in hcp metals*

Metal	K_1	K_2	η_1	η_2	γ
Mg	$\{10\bar{1}1\}$	$\{\bar{1}013\}$	$<10\bar{1}2>$	$<30\bar{3}2>$	
Mg	$\{10\bar{1}3\}$	$\{\bar{1}011\}$	$<\bar{3}032>$	$<\bar{1}012>$	$c/(\sqrt{3}a) - (3/4)\sqrt{3}a/c$
Zr, Ti	$\{11\bar{2}1\}$	(0001)	$<11\bar{2}\bar{6}>$	$<11\bar{2}0>$	a/c
Zr, Ti	$\{11\bar{2}2\}$	$\{11\bar{2}4\}$	$<\bar{1}\bar{1}23>$	$<\bar{2}\bar{2}43)$	$[c/a - 2a/c] = 0.224$ Zr,
and	$\{11\bar{2}4\}$	$\{11\bar{2}2\}$			$= .218$ for Ti

The atomic motions in twinning of the hcp lattice are very complex because the atoms do not all move in the direction of shear. Figure 11.13 shows the atom positions on adjacent $(\bar{1}2\bar{1}0)$ planes of a hypothetical hcp metal with a c/a ratio of $(11/12)\sqrt{3} = 1.588$. The atom motions are shown in Figure 11.13. Different atoms move different amounts and there are components of motion toward and away from the mirror plane. These atomic movements have been described as a *shear* plus a *shuffle* (Figure 11.14).

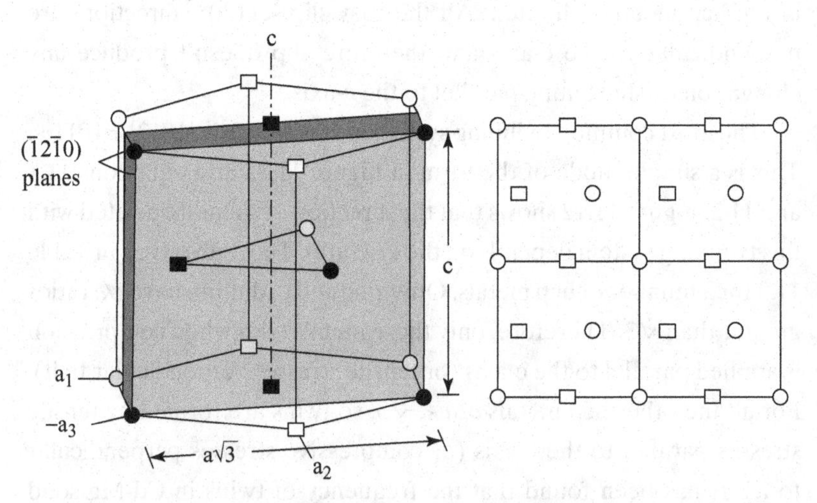

Figure 11.13. Plan view of the $(\bar{1}2\bar{1}0)$ plane showing the positions of two adjacent $(\bar{1}2\bar{1}0)$ planes. From W. F. Hosford, *The Mechanics of Crystals and Textured Polycrystals*, Oxford Sci. Publication (1993).

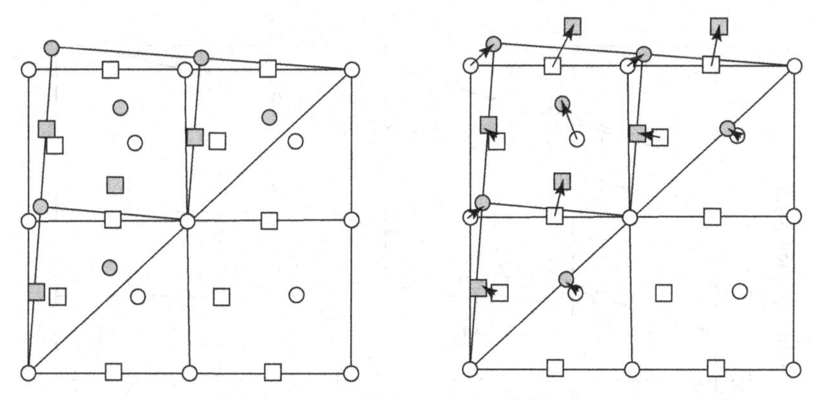

Figure 11.14. The atom movements on two adjacent $(\bar{1}2\bar{1}0)$ planes during $(10\bar{1}2)$ twinning. Note that the shear is not homogeneous. Some atoms move toward the twinning plane and some away From W. F. Hosford, *The Mechanics of Crystals and Textured Polycrystals,* Oxford Sci. Publication (1993).

There are other twinning modes for hcp metals that are not the simple type described by equations 11.1 and 11.2 and illustrated in Figure 11.15. Some of these are listed in Table 11.3. The twinning shear strains for the various modes depend on the c/a ratio as shown in Figure 11.16 but only changes sign for $\{10\bar{1}2\}<10\bar{1}1>$.

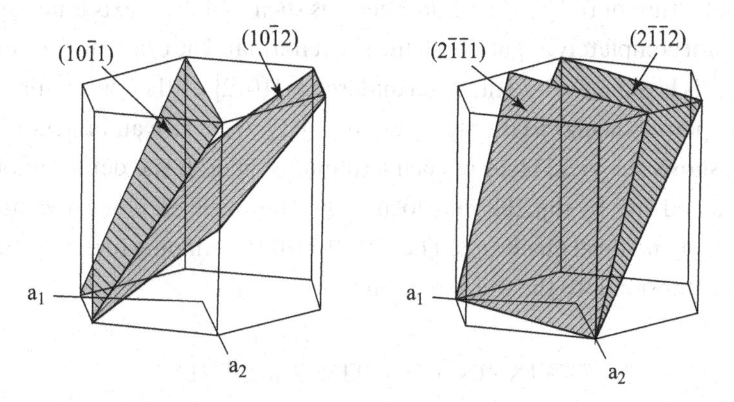

Figure 11.15. Twinning planes in hcp crystals. From W. F. Hosford, *The Mechanics of Crystals and Textured Polycrystals,* Oxford Sci. Publication (1993).

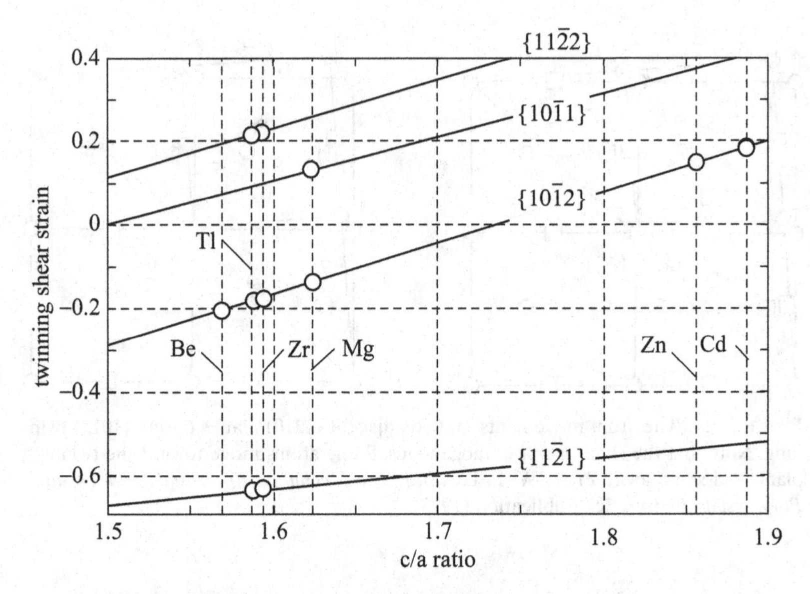

Figure 11.16. Variation of the shear strain, γ, for several twinning modes in hcp metals with *c/a* ratio. By convention, a positive shear strain causes elongation parallel to the c-axis. For $\{11\bar{2}1\}$ twinning, γ is negative, while for $\{11\bar{2}2\}$ and $\{10\bar{1}1\}$ twinning, γ is positive. For $\{10\bar{1}2\}$ twinning, the sign of γ depends on c/a. From W. F. Hosford, *The Mechanics of Crystals and Textured Polycrystals,* Oxford Sci. Publication (1993).

RESULTING ANISOTROPY

The texture of rolled a-titanium sheets is such that the c-axis is nearly, but not completely aligned with the sheet normal. The c/a ratio for titanium is less than $\sqrt{3}$, so titanium undergoes $\{10\bar{1}2\}<\bar{1}011>$ in compression perpendicular to the sheet. Figure 11.17, from Lee and Backofen [4], shows the yield locus of such a titanium sheet. If the deformation occurred only by slip, the yield locus would be an ellipse elongated into the first and third quadrants. The $\{10\bar{1}2\}<\bar{1}011>$ twinning foreshortens the extension into the third quadrant.

TWINNING IN OTHER METALS

Twinning modes in β-tin and crystal with structure are summarized in Table 11.4. The shears are not described by equation 11.1.

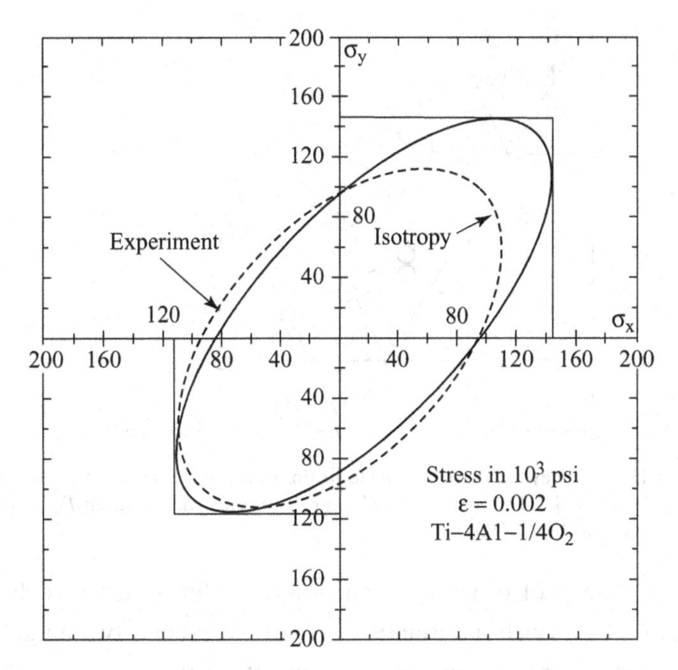

Figure 11.17. Yield locus for a sheet of Ti-4Al-1/4O. From D. Lee and W. A. Backofen, Trans TMS-AIME v. 242 (1968).

Figure 11.18 shows the lattice shear for {103}<$\bar{1}$03> twinning in tin.

PSEUDO-TWINNING

Twinning can occur in crystals that are random solid solutions. However, a shear strain that would cause a twin in a random solid solution

Table 11.4. *Twinning elements*

Metal	K1	K2	$\eta 1$	$\eta 2$.	γ
β-tin	{301}	{$\bar{1}$01}	<$\bar{1}$03>	<101>	(1/2)(a/c) – (3/2)c/a
	{101}	{$\bar{1}$01}	[10$\bar{1}$]	<110>	
rhombohedral (Sb, As, Bi)	{110}	(001)	[00$\bar{1}$]	<110>	$2\cos\alpha/[(1+\cos\alpha)/2-\cos^2\alpha]^{1/2}$ (where α is the angle between the axes)

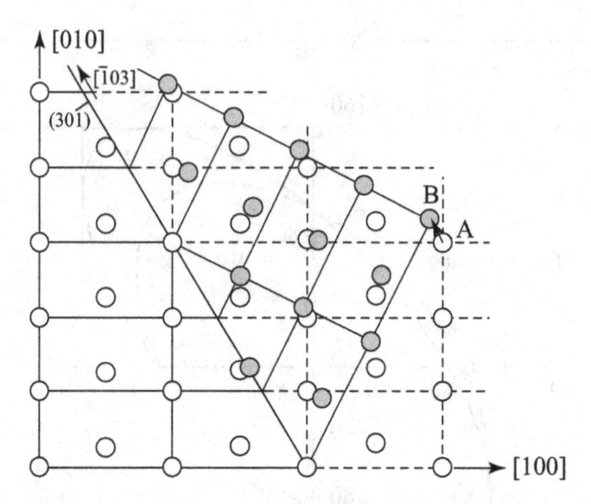

Figure 11.18. Plan view of the (010) plane in tin, showing the shear of {103} <1̄03> twinning. From W. F. Hosford, *The Mechanics of Crystals and Textured Polycrystals,* Oxford Sci. Publication (1993).

causes a change of ordering in an ordered solid solution as shown in Figure 11.19. Such deformations are called pseudo-twinning. The change nof order must increase the energy of the crystal, so there would be a tendency for the shear to reverse.

MARTENSITE TRANSFORMATION

Martensitic transformations are similar to mechanical twinning. They occur by sudden shearing of the lattice with atoms moving only a fraction of a normal interatomic distance. Originally the term *martensitic* transformation was reserved for steel, but now it is generally applied to all phase transformations that occur by shear without any composition change. Likewise the terms *austenite* and *martensite* are used to identify the high and low temperature phases. The extent of a martensitic transformation can be monitored be measuring changes of any property as shown in Figure 11.20. On cooling, the transformation starts at the M_s temperature and finished at the M_f temperature. On heating the martensite starts to revert to austenite at A_s and finishes at A_f. There is considerable hysteresis as illustrated in Figure 11.20. The A_s

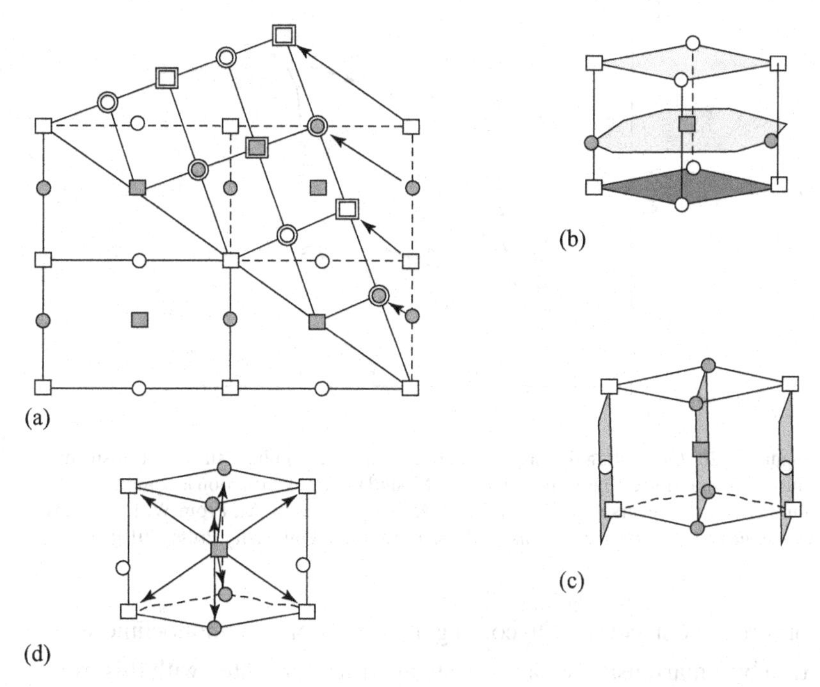

Figure 11.19. Change of order that accompanies pseudo twinning of an ordered bcc (B2) structure. Atoms of element A are shown as open squares and circles. The B atoms are indicated by filled squares and circles. The squares lie in one of the {110} planes and the circles in the other. The atom movements shown in (a) change the original order of alternating {001} planes (b) to alternating {110} planes (c). After the shear, each B atom has 4 A nearest neighbors and 4 B atoms as nearest neighbors instead of 8 A atom nearest neighbors. From W. F. Hosford, *The Mechanics of Crystals and Textured Polycrystals,* Oxford Sci. Publication (1993).

and A_f temperatures are higher than the M_s and M_f temperatures. Between the A_s and M_s temperatures, martensite can be formed by deformation.

SHAPE MEMORY

This is an effect in which plastic deformation (or at least what appears to be plastic deformation) is reversed on heating. The alloys that exhibit this effect are invariably ordered solid solutions that undergo a martensitic transformation on cooling. The alloy TiNi at 200°C has an

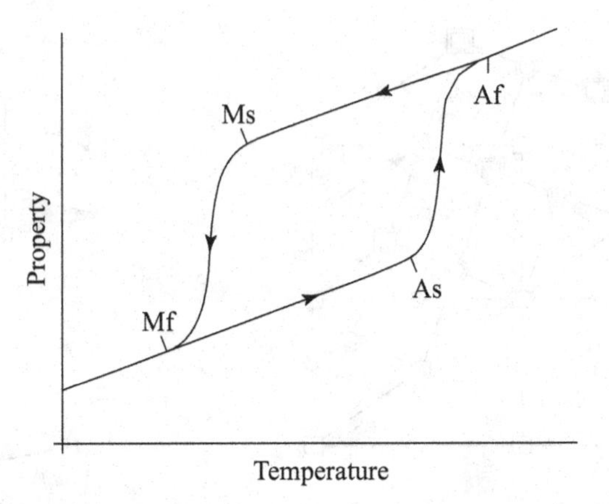

Figure 11.20. On cooling the high temperature (austenite) phase starts to transform to the low temperature phase (martensite) at Mf and the transformation is complete at Mf. On heating the reverse reaction starts at As and finishes at Af. From W. F. Hosford, *Mechanical Behavior of Materials*, 2nd ed., Cambridge University Press (2010).

ordered bcc structure. On cooling, it transforms to a monoclinic structure by a martensitic shear. The shear strain associated with this transformation is about 12 percent. If only one variant of the martensite were formed, the strain in the neighboring untransformed lattice would be far too high to accommodate. To decrease this compatibility strain, two mirror image variants form in such a way that there is no macroscopic strain. The macroscopic shape is the same as before the transformation. Figure 11.21 illustrates this. The boundaries between the two variants are highly mobile so if the resulting structure is deformed, the deformation is easily accommodated by movement of the boundaries in what appears to be plastic deformation. Figure 11.22 shows a stress strain curve. Heating the deformed material above the Af temperature causes it to transform back to the ordered cubic structure by martensitic shear. For both variants, the martensite shears must be of the correct sign to restore the correct order. The overall effect is that the deformation imposed on the low temperature martensitic form is reversed. The critical temperatures for reversal in TiNi are typically in the range of 80 to 100°C but are sensitive to very minor changes in

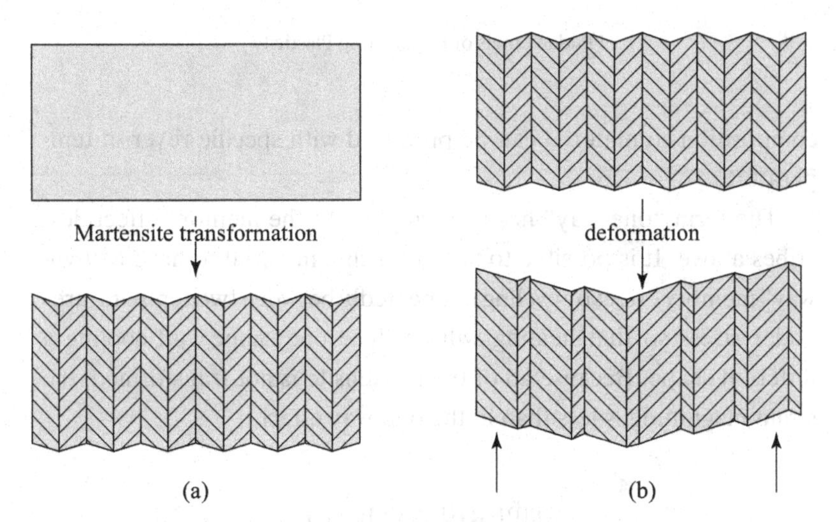

Figure 11.21. Schematic a. As the material is cooled it undergoes a martensitic transformation. By transforming to equal amounts of two variants, the macroscopic shape is retained. b. Deformation occurs by movement of variant boundaries so more the more favorably oriented variant grows at the expense of the other. Adapted from a sketch by D. Grummon. From W. F. Hosford, *Mechanical Behavior of Materials*, 2nd ed., Cambridge University Press (2010).

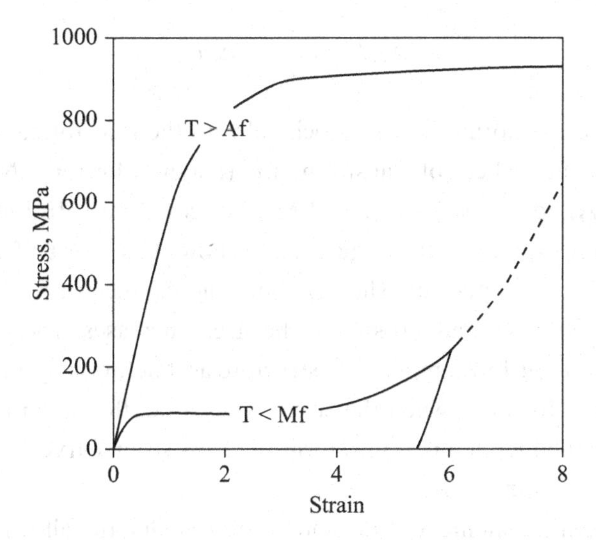

Figure 11.22. Stress strain curve for a shape memory material. The lower curve is for deformation when the material is entirely martensitic. The deformation occurs by movement of variant boundaries. After all of the material is of one variant, the stress rises rapidly. The upper curve is for the material above its Af temperature. Adapted from a sketch by Grummon W. F. Hosford, *Mechanical Behavior of Materials*, 2nd ed., Cambridge University Press (2010).

composition so material can be produced with specific reversal temperatures.

The term "one-way" has been applied to the memory effect describes above. It is possible to "train" such a material to have a "two-way" memory effect by cycling it repeatedly between two temperatures under stress so that heating will produce one shape and cooling a different shape. Because all of the martensite is not transformed, the strain magnitude is less than in the one-way effect.

SUPERELASTICITY

This phenomenon is closely related to the shape memory effect except that the deformation temperature is above the normal A_f temperature. However, the A_f temperature is raised by applied stress. According to the Clausius-Clapyron equation,

$$d(A_f)/d\sigma = T\varepsilon_0/\Delta H, \qquad 11.3$$

where ε_0 is the normal strain associated with the transformation and ΔH is the latent heat of transformation (about 20J/g for TiNi). The terms $d(A_s)/d\sigma$, $d(M_s)/d\sigma$, and $d(M_f)/d\sigma$, could be substituted for $d(A_f)/d\sigma$ in equation 10.3. Figure 10.23 shows this effect. If a stress is applied at a temperature slightly above the A_f, the A_f, A_s, M_s, and M_f temperatures all are crossed as the stress increases. The material transforms to its low temperature structure as it deforms by a martensitic shear. However, when the stress is released, the material again reverts to the high temperature form. A stress strain curve for Fe_3Be is shown in Figure 11.24.

For both the memory effect and superelasticity, the alloy must be ordered, there must be a martensitic transformation and the variant boundaries must be mobile. The difference between the shape memory effect and superelasticity is shown schematically in Figure 11.25. For the super-elastic effect, deformation starts when the material is

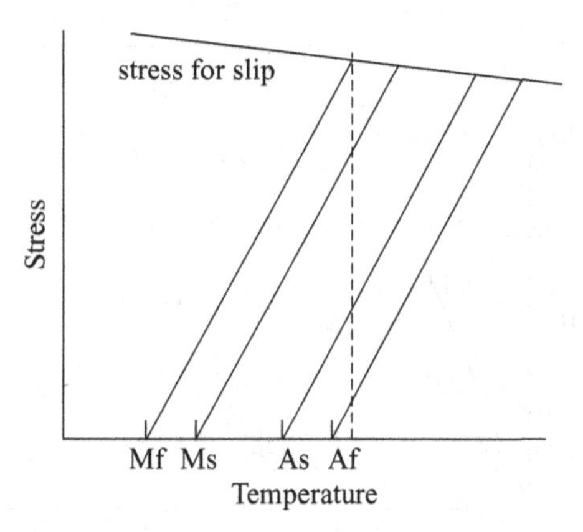

Figure 11.23. As stress is applied to a super- elastic material, the A_f, A_s, M_s, and M_f temperatures for the material undergoes martensitic shear strains. When the stress is removed, the material reverts to its high temperature form reversing all of the martensitic deformation. After a sketch by Grummon. W. F. Hosford, *Mechanical Behavior of Materials*, 2nd ed., Cambridge University Press (2010).

austenitic whereas for the shape memory effect the martensite phase must be deformed.

NOTES OF INTEREST

For many years it was widely thought that annealing twins were common in fcc metals but not in bcc or hcp metals, and that mechanical twinning may occur in bcc and hcp metals but never in fcc metals. The reasons for this inverse relationship were not explained. However, it was so strongly believed that when in 1957, Blewitt, Coltman, and Redman first reported mechanical twinning in copper at very low temperatures [8], other workers were very skeptical until these findings were confirmed by other researchers.

In 1932, Swedish researcher Arne Olader first observed the shape-memory effect in a gold-cadmium alloy. Other shape-memory alloys include CuSn, InTi, TiNi, and MnCu. TiNi. The superplastic effect in TiNi was first found by William Buehler and Frederick Wang in 1962

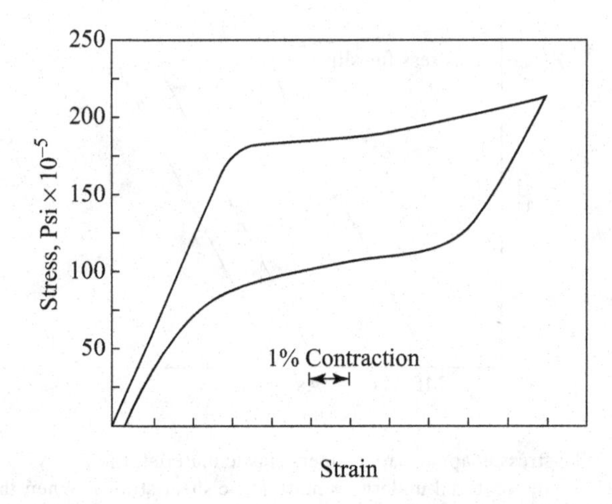

Figure 11.24. The stress-strain curve for superelastic Fe_3Be. After the initial Hookian strain, the material deforms by martensitic transformation. On unloading the reverse martensitic transformation occurs at a lower stress. From R. H. Richman in *Deformation Twinning*, TMS-AIME (1963).

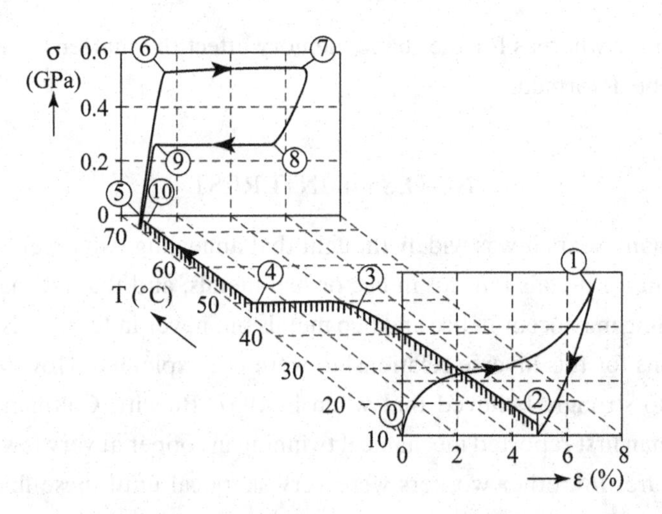

Figure 11.25. Schematic illustration of the difference between shape memory and superelastic effects. For shape memory, the deformation occurs at a temperature for which the material is martensitic. A superelastic effect occurs when the deformation occurs just above the A_f temperature. From J. A. Shaw, *Int. J. of Plasticity*, v. 16 (2000) (2000).

at the Naval Ordnance Laboratory. They called the alloy Nitinol from after Nickel Titanium Naval Ordnance Laboratory.

REFERENCE

1. T. H. Blewitt, R. R. Coltman, and J. K. Redman. *J. Applied Physics*, v. 28 (1957).

GENERAL REFERENCES

1. E. O. Hall, *Twinning and Diffusionless Transformations* in Metals, Butterworths (1954).
2. *Deformation Twinning*, R. E Reed-Hill, J. P Hirth, and H. C., Rogers, eds. TMS-AIME (1963).
3. W. F. Hosford, *The Mechanics of Crystals and Textured Polycrystals*, Oxford University Press (1993).

12

EFFECTS OF STRAIN HARDENING AND STRAIN-RATE DEPENDENCE

The terms *strain hardening* and *work hardening* are used interchangeably to describe the increase of the stress level necessary for continued plastic deformation. The term *flow stress* is used to describe the stress necessary to continue deformation at any stage of plastic strain. Prediction of the energy absorption in automobile crashes and the stresses around cracks require mathematical descriptions of true stress-strain curves. They are also required in design of dies for stamping parts. Various approximations are possible. Which approximation is best depends on the material, the nature of the problem, and the need for accuracy.

MATHEMATICAL APPROXIMATIONS

The simplest model is *ideal plasticity* in which strain hardening is ignored. The flow stress, σ, is independent of strain, so

$$\sigma = Y, \qquad 12.1$$

where Y is the tensile yield strength (see Figure 12.1a). For linear strain hardening (Figure 12.1b).

$$\sigma = Y + A\varepsilon. \qquad 12.2$$

It is more common for materials to strain harden with a hardening rate that decreases with strain. For many metals a log-log plot of

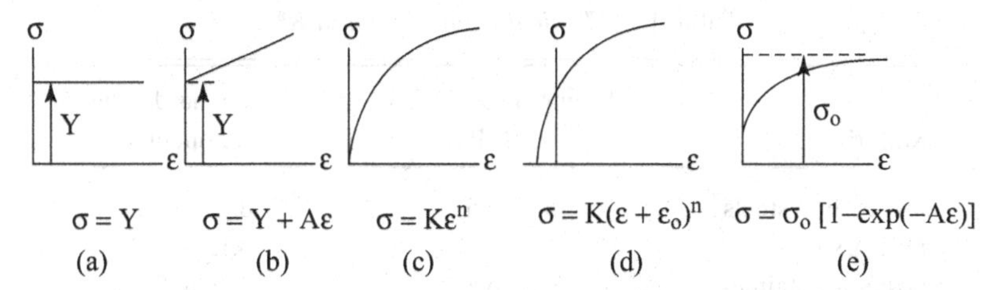

Figure 12.1. Mathematical approximations to true stress strain curves. From W. F. Hosford, *Mechanical Behavior of Materials*, 2nd ed., Cambridge University Press (2011).

true stress versus true strain is nearly linear [1]. In this case, a power law,

$$\sigma = K\varepsilon^n, \tag{12.3}$$

is a reasonable approximation (Figure 12.1c). A better fit is often obtained with

$$\sigma = K(\varepsilon + \varepsilon_o)^n. \tag{12.4}$$

(see Figure 12.1d). This expression is useful where the material has undergone a pre-strain of εo. Still another model is a saturation model suggested by Voce [2] (Figure 12.1e) is

$$\sigma = \sigma_o[1 - \exp(-A\varepsilon)]. \tag{12.5}$$

Equation 12.5 predicts that the flow stress approaches an asymptote, σ_o at high strains. This model seems to be reasonable for a number of aluminum alloys.

POWER-LAW APPROXIMATION

The most commonly used expression is the simple power law (equation 12.3). Typical values of the exponent n are in the range of 0.1 to 0.6. Table 12.1 lists K and n for various materials. As a rule, high strength materials have lower n-values than low strength materials. Figure 12.2 shows that the exponent, n, is a measure of the persistence

Table 12.1 *Typical values of n and K*[a]

Material	Strength coefficient, K (MPa)	Strain hardening exponent, n
low-carbon steels	525 to 575	0.20 to 0.23
HSLA steels	650 to 900	0.15 to 0.18
austenitic stainless	400 to 500	0.40 to 0.55
copper	420 to 480	0.35 to 0.50
70/30 brass	525 to 750	0.45 to 0.60
aluminum alloys	400 to 550	0.20 to 0.30

[a] From various sources including ref. 3.

of hardening. If n is low, the work hardening rate is initially high but the rate decreases rapidly with strain. On the other hand, with a high n, the initial work hardening is less rapid but continues to high strains. If $\sigma = K\varepsilon^n$, $\ln\sigma = \ln K + n\ln\varepsilon$ so the true stress strain relation plots as a straight line on log-log coordinates as shown in Figure 12.3. The

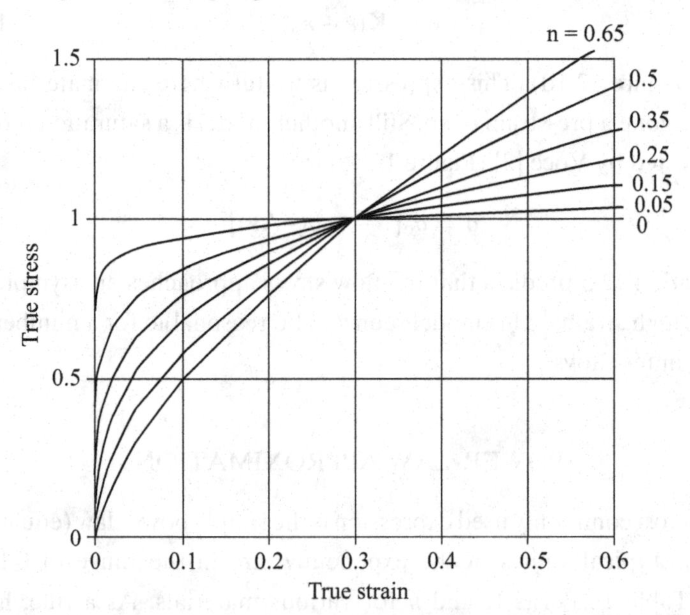

Figure 12.2. True stress – strain curves for $\sigma = K\varepsilon^n$ with several values of n. To make the effect of n on the shape of the curves apparent, the value of K for each curve has been adjusted so that it passes through $\sigma = 1$ at $\varepsilon = 0.3$. From W. F. Hosford, *The Mechanics of Crystals and Textured Polycrystals*, Oxford Sci. Publication (1993).

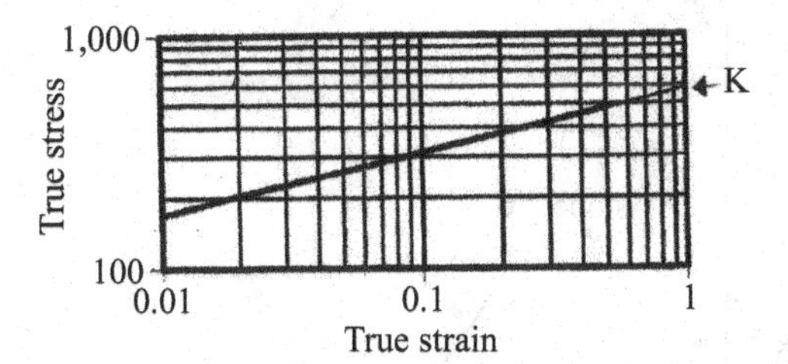

Figure 12.3. A plot of the true stress strain curve on logarithmic scales. The straight line indicates that $\sigma = k\varepsilon^n$. The slope $= n$ and K = the intercept at $\varepsilon = 1$. From W. F. Hosford, *Mechanical Behavior of Materials*, 2nd ed., Cambridge University Press (2011).

exponent, n, is the slope of the line. The pre-exponential, K, can be found by extrapolating to $\varepsilon = 1.0$. K is the value of σ at this point. The exponent n is important in stretch forming because it indicates the ability of a metal to distribute the straining over a wide region. Often a log-log plot of the true stress strain curve deviates from linearity at low or high strains. In such cases, it is still convenient to use equation 12.3 over the strain range of concern. The value of n is then taken as the slope of the linear portion of the curve.

$$n = \mathrm{d}(\ln\sigma)/\mathrm{d}(\ln\varepsilon) = (\varepsilon/\sigma)\mathrm{d}\sigma/\mathrm{d}\varepsilon. \qquad 12.6$$

NECKING

As a tensile specimen is extended, the level of true stress, σ, rises but the cross-sectional area carrying the load decreases. The maximum load-carrying capacity, $F = \sigma A$, is reached when $\mathrm{d}F = 0$. Differentiating gives

$$A\mathrm{d}\sigma + \sigma \mathrm{d}A = 0. \qquad 12.7$$

Since the volume, AL, is constant, $\mathrm{d}A/A = -\mathrm{d}L/L = -\mathrm{d}\varepsilon$. Rearranging terms $\mathrm{d}\sigma = -\sigma\mathrm{d}A/A = \sigma\mathrm{d}\varepsilon$, or

$$\mathrm{d}\sigma/\mathrm{d}\varepsilon = \sigma. \qquad 12.8$$

Equation 12.8 simply states that the maximum load is reached when the rate of work hardening is numerically equal to the stress level.

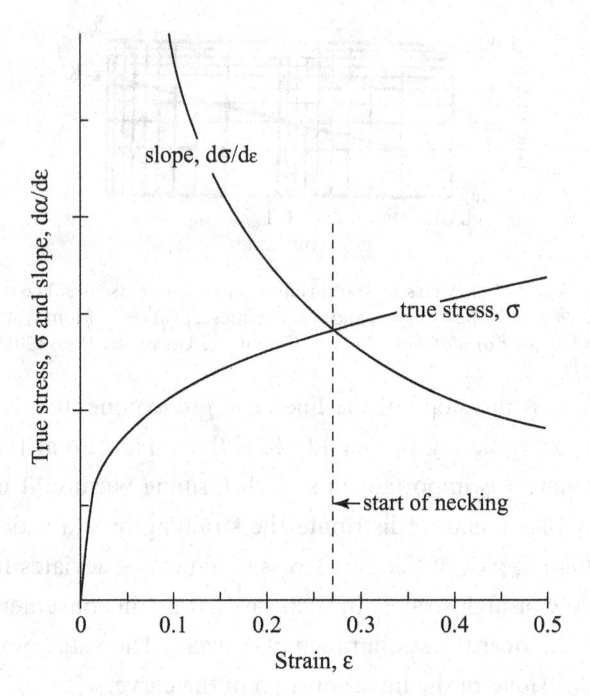

Figure 12.4. The condition for necking in a tension test is met when the true stress, σ, equals the slope, $d\sigma/d\varepsilon$, of the true stress-strain curve. From W. F. Hosford, *Mechanical Behavior of Materials*, 2nd ed., Cambridge University Press (2011).

As long as $d\sigma/d\varepsilon > \sigma$, deformation occurs uniformly along the test bar. However, once the maximum load is reached, the deformation localizes. Any region that deforms even slightly more than the others has a lower load-carrying capacity and the load drops to that level. Other regions will cease to deform so deformation will localize into a neck. Figure 12.4 is a graphical illustration.

If a mathematical expression is assumed for the stress-strain relationship, the limit of uniform elongation can be found analytically. For example, with power-law hardening, equation 12.3,

$\sigma = K\varepsilon^n$, and $d\sigma/d\varepsilon = nK\varepsilon^{n-1}$. Combining equations 12.3 and 12.8, $K\varepsilon^n = nK\varepsilon^{n-1}$, which simplifies to

$$\varepsilon = n \tag{12.9}$$

so the strain at the start of necking equals the strain-hardening exponent, n. Uniform elongation in a tension test occurs before necking. Therefore, so materials with a high n value have large uniform elongations.

Because the tensile strength is the engineering stress at maximum load, the power-law hardening rule can be used to predict it. Substituting the strain, n, into, into equation 12.3, the true stress at maximum load becomes,

$$\sigma_{\text{max load}} = \mathrm{K}n^n. \qquad 12.10$$

Next substituting $\sigma = \sigma/(1+\varepsilon) = \sigma\exp(-\varepsilon) = \sigma\exp(-n)$ into equation 12.10, the tensile strength is

$$\sigma_{\max} = \mathrm{K}n^n\exp(-n) = \mathrm{K}(n/e)^n, \qquad 12.11$$

where e is the base of natural logarithms.

Similarly, the uniform elongation and tensile strength may be found for other approximations to the true stress-strain curve.

WORK PER VOLUME

The area under the true stress strain curve is the work per volume, w, expended in the deforming a material. That is, $w = \int \sigma d\varepsilon$. With power-law hardening,

$$w = \mathrm{K}\varepsilon^{n+1}/(n+1). \qquad 12.12$$

EFFECT OF STRAIN HARDENING ON YIELD LOCI

According to the *isotropic hardening* model, the effect of strain hardening is simply to expand the yield locus without changing its shape. The stresses for yielding are increased by the same factor along all loading paths. This is the basic assumption that $\bar{\sigma} = \mathrm{f}(\bar{\varepsilon})$. The isotropic

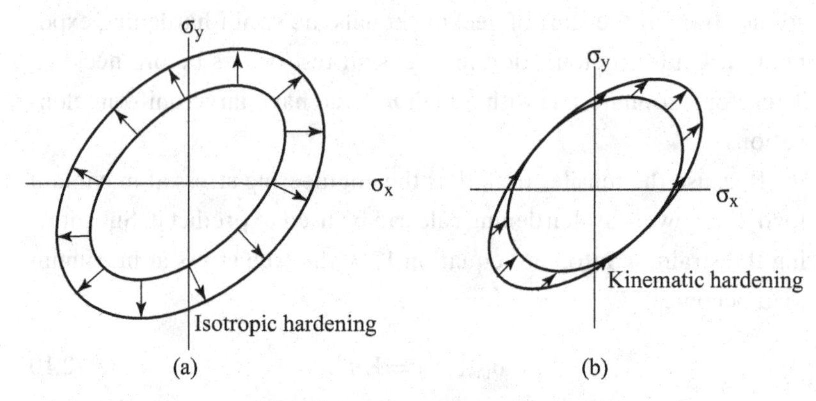

Figure 12.5. The effect of strain hardening on the yield locus. The isotropic model (a) predicts an expansion of the locus. The kinematic hardening model predicts a translation of the locus in the direction of the loading path. From W. F. Hosford, *Mechanical Behavior of Materials*, 2nd ed., Cambridge University Press (2011).

hardening model can be applied to anisotropic materials. It does not imply that the material is isotropic.

An alternative model is *kinematic hardening*. According to this model, plastic deformation simply shifts the yield locus in the direction of the loading path without changing its shape or size. If the shift is large enough, unloading may actually cause plastic deformation. The kinematic model is probably better for describing small strains after a change in load path. However, the isotropic model is better for describing behavior during large strains after a change of strain path. Figure 12.5 illustrates both models.

STRAIN-RATE DEPENDENCE OF FLOW STRESS

The flow stress of most materials rise with increased strain rates. The amount of the rise depends on the material and its temperature. For most metallic materials, the effect near room temperature is so small that it is often neglected. A factor of ten increase of strain rate may raise the level of the stress strain curve by only 1 or 2 percent. At elevated temperatures, however, the effect of strain rate on flow stress is much greater. Increasing the strain rate by a factor of ten may raise the stress-strain curve by 50 percent or more.

There is a close coupling of the effects of temperature and strain rate on the flow stress. Increased temperatures have the same effects as deceased temperatures. This coupling can be understood in terms of the Arrhenius rate equation.

The average strain rate during most tensile tests is in the range of 10^{-3} to 10^{-2}/s. If it takes 5 minute during the tensile test to reach a strain of 0.3, the average strain rate is $\dot{\varepsilon} = 0.3/(5 \times 60) = 10^{-3}/\text{s}$. A strain rate of $\dot{\varepsilon} = 10^{-2}/\text{s}$ a strain of 0.3 will occur in 30 seconds. For many materials the effect of the strain rate on the flow stress, σ, at a fixed strain and temperature can be described by a power-law expression

$$\sigma = C\dot{\varepsilon}^m, \qquad 12.13$$

where the exponent, m, is called the strain-rate sensitivity. The relative levels of stress at two strain rates (measured at the same total strain) is given by

$$\sigma_2/\sigma_1 = (\dot{\varepsilon}_2/\dot{\varepsilon}_1)^m, \qquad 12.14$$

or $\ln(\sigma_2/\sigma_1) = m\ln(\dot{\varepsilon}_2/\dot{\varepsilon}_1)$. If σ_2 is not much greater than σ_1,

$$\ln(\sigma_2/\sigma_1) \approx \Delta\sigma/\sigma \qquad 12.15$$

Equation 12.14 can be simplified to

$$\Delta\sigma/\sigma \approx m\ln(\dot{\varepsilon}_2/\dot{\varepsilon}_1) = 2.3m\log(\dot{\varepsilon}_2/\dot{\varepsilon}_1). \qquad 12.16$$

At room temperature, the values of m for most engineering metals are between -0.005 and $+0.015$ as shown in Table 12.2.

Consider the effect of a ten-fold increase in strain rate, ($\dot{\varepsilon}_2/\dot{\varepsilon}_1 = 10$) with $m = 0.01$. Equation 12.16 predicts that the level of the stress increases by only $\Delta\sigma/\sigma = 2.3(0.01)(1) = 2.3\%$. This increase is typical of room temperature tensile testing. It is so small that the effect of strain-rate is often ignored. A plot of equation 12.14 in Figure 12.6

Table 12.2 *Typical values of the strain-rate exponent, m, at room temperature*

Material	*m*
low-carbon steels	0.010 to 0.015
HSLA steels	0.005 to 0.010
austenitic stainless steels	−0.005 to +0.005
ferritic stainless steels	0.010 to 0.015
copper	0.005
70/30 brass	−0.005 to 0
aluminum alloys	−0.005 to +0.005
α-titanium alloys	0.01 to 0.02
zinc alloys	0.05 to 0.08

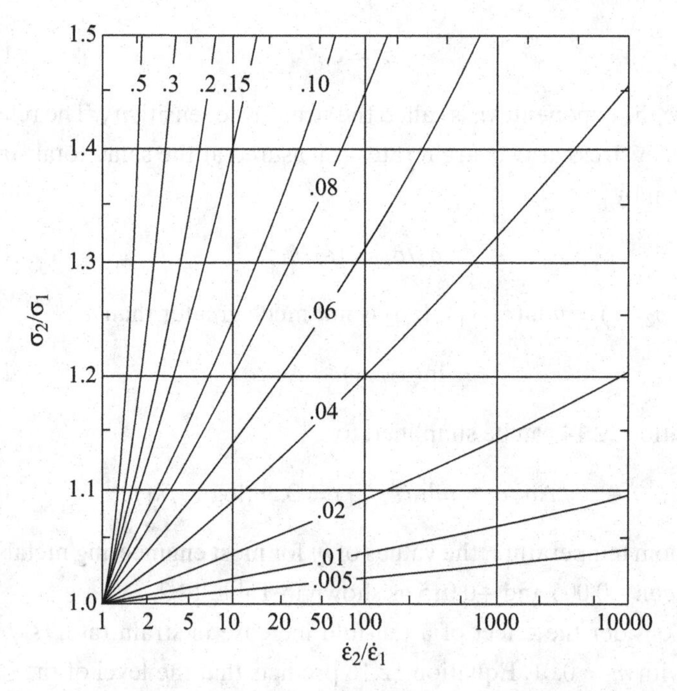

Figure 12.6. The dependence of flow stress on strain rate for several values of the strain-rate sensitivity, m, according to equation 12.14. From W. F. Hosford and R. M. Caddell, *Metal Forming; Mechanics and Metallurgy*, 4th ed., Cambridge University Press (2011).

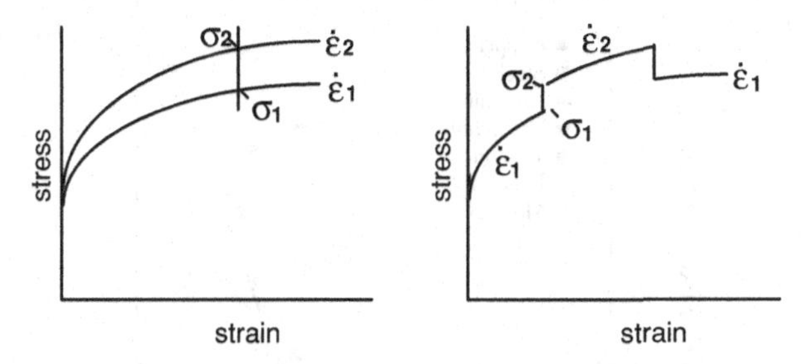

Figure 12.7. Two methods of determining the strain-rate sensitivity. Either continuous stress strain curves at different strain rates can be compared at the same strain (left) or sudden changes of strain rate can be made and the stress levels just before and just after the change compared (right). In both cases, equation 11.16 can be used to find m. In rate-change tests, $(\dot{\varepsilon}_2/\dot{\varepsilon}_1)$ is typically 10 or 100. From W. F. Hosford and R. M. Caddell, *Metal Forming; Mechanics and Metallurgy*, 4th ed., Cambridge University Press (2011).

shows how the relative flow stress depends on strain rate for several levels of m. The increase of flow stress, $\Delta\sigma/\sigma$, is small unless either m or $(\dot{\varepsilon}_2/\dot{\varepsilon}_1)$ is high.

Two ways of determining the value of m are illustrated in Figure 12.7. One method is to compare the levels of stress at the some fixed strain in two continuous tensile tests run at different strain rates The other way is to suddenly change the strain rate during a test and compare the levels of stress immediately before and after the change. The latter method is easier and therefore more common. The two methods may give somewhat different values for m. In both cases, equation 12.16 can be used to find m. In rate-change tests, $(\dot{\varepsilon}_2/\dot{\varepsilon}_1)$ is typically 10 or 100.

For most metals, the value of m for most metals is low near room temperature but increases with temperature rising rapidly above half of the melting point ($T > T_m/2$) on an absolute temperature scale. In special cases, m may reach 0.5 or higher. Figure 12.9 shows the temperature dependence of m for several metals. For some alloys, there is a minimum between $0.2T_m$ and $0.3T_m$. For aluminum alloy 2024, (Figure 12.10) the rate sensitivity is slightly negative in this temperature range.

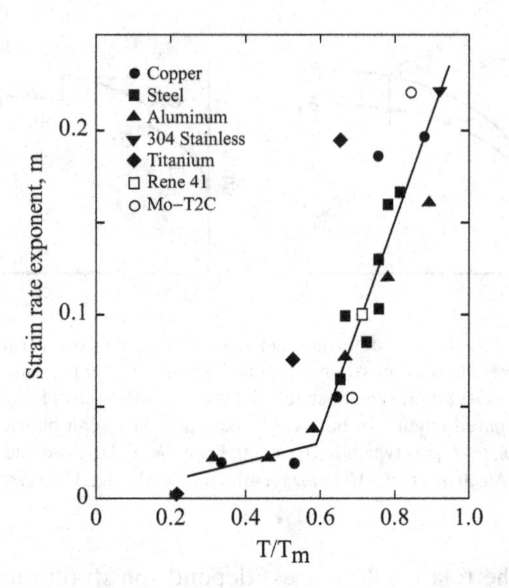

Figure 12.8. Variation of the strain-rate sensitivity, *m*, with temperature for several metals. Above about half of the melting point, m rises rapidly with temperature. From W. F. Hosford and R. M. Caddell, *Metal Forming; Mechanics and Metallurgy*, 4th ed., Cambridge University Press (2011).

For values of *m* above about 0.2, necking in tension is very gradual. This leads to very high elongations as indicated in Figure 12.9. The term *superplasticity* has been used to describe this phenomenon.

For body-centered cubic metals, a better description of strain-rate dependence is

$$\sigma = \mathrm{C} + m'\ln\varepsilon \qquad 12.17$$

where C is the flow stress at a reference strain rate and m' is the rate sensitivity. With this expression, a change of strain rates from ε_1 to ε_2 will change the flow stress by (See Figure 12.10)

$$\Delta\sigma = m'\ln(\varepsilon_1/\varepsilon_2) \qquad 12.18$$

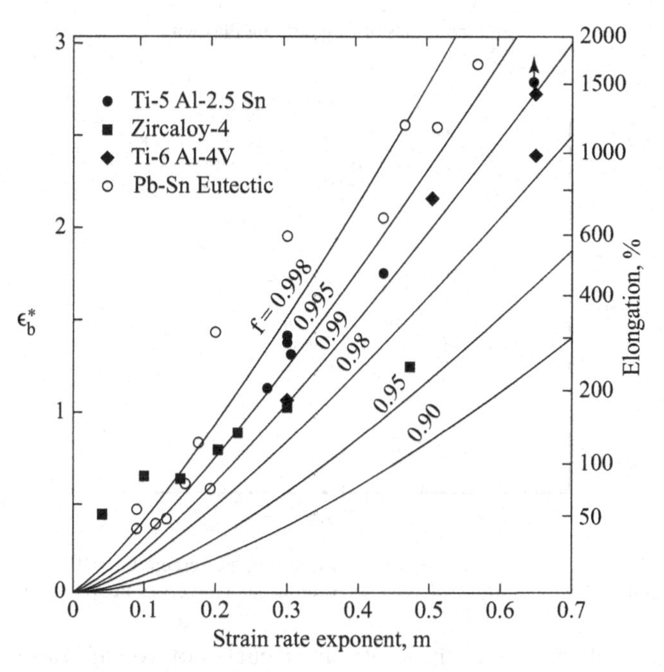

Figure 12.9. Limiting strains, ε_b^*, in unreduced sections of stepped tensile specimens as a function of m and f. Values of percent elongation corresponding to ε_b^* are indicated on the right. From W. F. Hosford and R. M. Caddell, *Metal Forming: Mechanics and Metallurgy*, 4th ed., Cambridge University Press (2011).

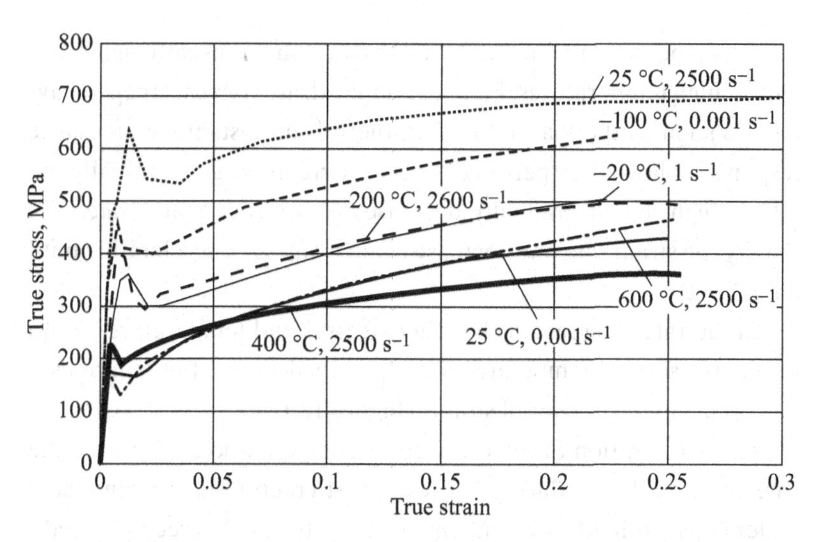

Figure 12.10. Stress-strain curves for iron at 25° C. Note that the difference in the level of the curves is independent of the stress level. From G. T. Gray in *ASM Metals Handbook*, v 8, (2005).

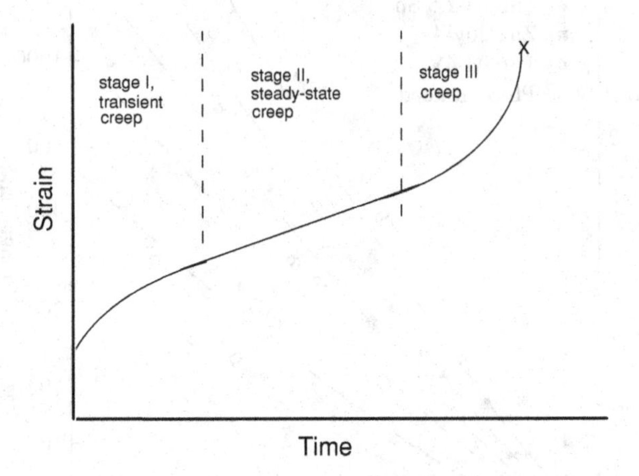

Figure 12.11. Typical creep curve showing three stages of creep. From W. F. Hosford, *Mechanical Behavior of Materials*, 2nd ed., Cambridge University Press (2011).

Figure 12.11 shows that the stress-strain curves of iron are raised by a constant level, $\Delta\sigma$, that is independent of the stress level.

CREEP

Creep is time-dependent plastic deformation that is usually significant only at high temperatures. Figure 12.11 illustrates typical creep behavior. As soon as the load is applied, there is an instantaneous elastic response, followed by period of transient creep (*stage I*). Initially the rate is high, but it gradually decreases to a steady state (*stage II*). Finally the strain rate may increase again (*stage III*), accelerating until failure occurs.

Creep rates increase with higher stresses and temperatures. With lower stresses and temperatures creep rates decrease but failure usually occurs at lower overall strains (Figure 12.12).

The acceleration of the creep rate in stage III occurs because the true stress increases during the test. Most creep tests are conducted under constant load (constant engineering stress). As creep proceeds, the cross-sectional area decreases so the true stress increases. Porosity develops in the later stages of creep, further decreasing the load-bearing cross section.

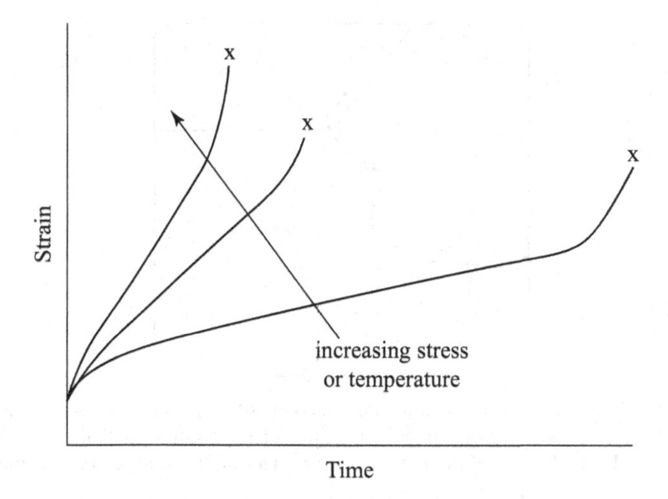

Figure 12.12. Decreasing temperature and stress lead to slower creep rates, but failure often occurs at a lower strains. From W. F. Hosford, *Mechanical Behavior of Materials*, 2nd ed., Cambridge University Press (2011).

CREEP MECHANISMS

Several mechanisms may contribute to creep. These include viscous flow, diffusional flow, and dislocation movement. Viscous flow is the dominant mechanism in amorphous materials. With Newtonian viscosity, the rate of strain, $\dot{\gamma}$, is proportional to the stress, τ,

$$\dot{\gamma} = \tau/\eta, \quad 12.19$$

where is the viscosity. For tensile deformation, this may be expressed as

$$\dot{\varepsilon} = \sigma/\eta', \quad 12.20$$

where $\eta' = 3\,\eta$.

In polycrystalline materials, grain-boundary sliding is viscous in nature. The sliding velocity on the boundary is proportional to the stress and inversely proportional to the viscosity, η. The rate of extension, $\dot{\varepsilon} = C(\sigma/\eta)/d$, depends on the amount of grain boundary area per volume and is therefore inversely proportional to the grain size, d.

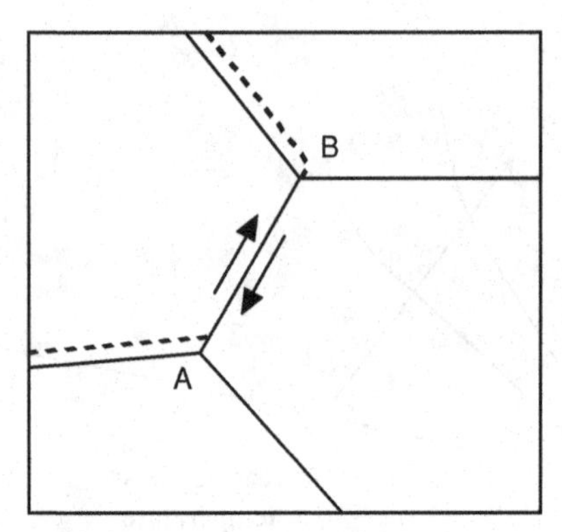

Figure 12.13. Grain boundary sliding causes incompatibilities at both ends of the planes, A and B, on which sliding occurs. This must be relieved by another mechanism for sliding to continue. From W. F. Hosford, *Mechanical Behavior of Materials*, 2nd ed., Cambridge University Press (2011).

Viscous flow is thermally activated, so $\eta = \eta_o \exp[Qv/(RT)]$. The strain-rate attributable to grain-boundary sliding can be written as,

$$\dot{\varepsilon}_v = A_V(\sigma/d)\exp(-Q\Delta\sigma/RT). \qquad 12.21$$

If grain boundary sliding were the only active mechanism, there would be an accumulation of material at one end of each boundary on which sliding occurs and a deficit at the other end, as sketched in Figure 12.13. This incompatibility must be relieved by another deformation mechanism, one involving dislocation motion, diffusion, or grain boundary migration. Figure 12.14 shows grain boundary sliding in aluminum.

Diffusion-controlled creep: A tensile stress increases the separation of atoms on grain boundaries that are normal to the stress axis, and the lateral contraction decreases the separation of atoms on grain boundaries that are parallel to the stress axis. The result is a driving force for diffusional transport of atoms from grain boundaries parallel to the tensile stress to boundaries normal to the tensile stress. Such diffusion produces a plastic elongation as shown in Figure 12.15. The specimen elongates as atoms are added to grain boundaries perpendicular to the stress.

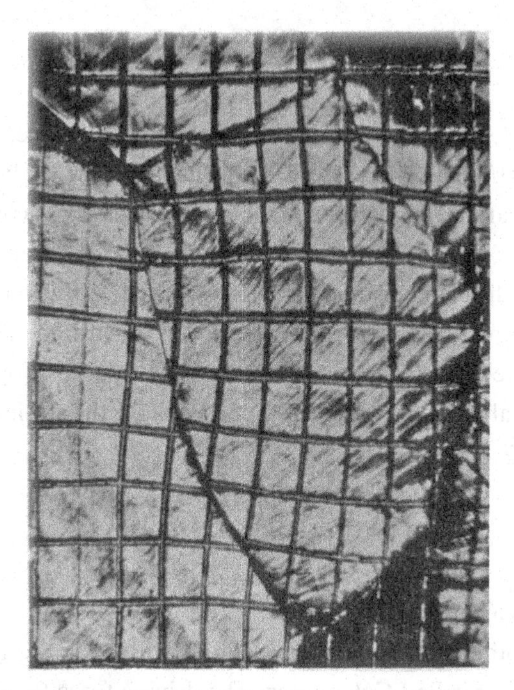

Figure 12.14. Grain boundary sliding in high-purity aluminum. From H. Brunner, *ScD Thesis*, MIT 1957.

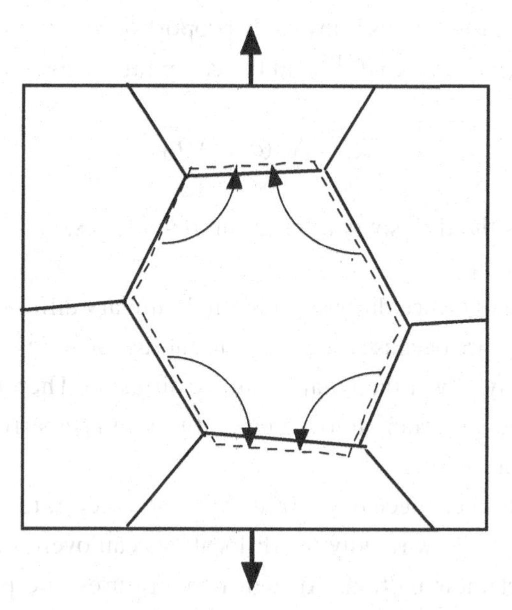

Figure 12.15. Creep by diffusion between grain boundaries. As atoms diffuse from lateral boundaries to boundaries normal to the tensile stress, the grain elongates and contracts laterally. From W. F. Hosford, *Mechanical Behavior of Materials*, 2nd ed., Cambridge University Press (2011).

If the creep occurs by diffusion through the lattice, it is called *Nabarro-Herring* [3, 4] *creep*. The diffusional flux, J, between the boundaries parallel and perpendicular to the stress axis is proportional to the stress, σ, and the lattice diffusivity, D_L, and it is inversely proportional to the diffusion distance, $d/2$, between the diffusion source and sink. Therefore, $J = CD_L\sigma/(d/2)$ where C is a constant. The velocity, v, at which the diffusion source and sink move apart is proportional to the diffusional flux, so $v = CD_L\sigma/(d/2)$. Because the strain rate equals $v/(d/2)$,

$$\dot{\varepsilon}_{N-H} = A_L(\sigma/d^2)D_L \qquad 12.22$$

where, A_L is a constant.

On the other hand, if creep occurs by diffusion along the grain boundaries, it is called *Coble creep*. The driving force for Coble creep is the same as for Nabarro-Herring creep. The total number of grain-boundary diffusion paths is inversely proportional to the grain size, so now J is proportional to $d^{-1/3}$ and the creep rate is given by

$$\dot{\varepsilon}_C = A_G(\sigma/d^3)D_{gb}, \qquad 12.23$$

where D_{gb} is the diffusivity along grain boundaries, and A_G is a constant.

The ratio of lattice diffusion to grain boundary diffusion increases with temperature because the activation energy for grain boundary diffusion is always lower than that for lattice diffusion. Therefore, Coble-creep is more important at low temperatures and Nabarro-Herring at high temperatures.

Slip is another mechanism of creep. The creep rate, in this case, is controlled by how rapidly the dislocations can overcome obstacles that obstruct their motion. At high temperatures, the predominant mechanism for overcoming obstacles is dislocation climb [5] (Figure 12.16). With climb, the creep rate is not dependent on grain size, but

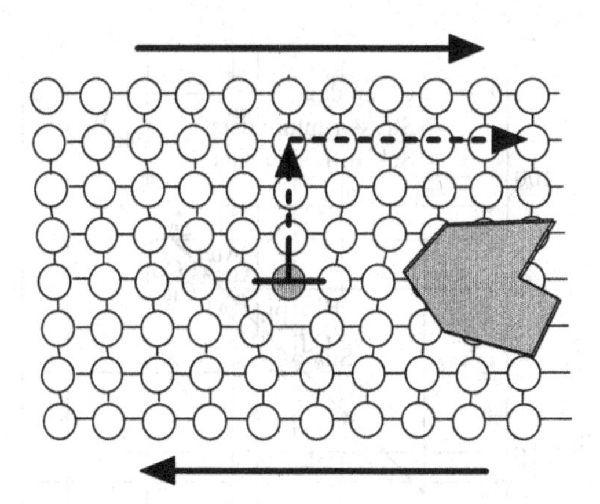

Figure 12.16. Climb-controlled creep. An edge dislocation can climb by diffusion of atoms away from the dislocation (vacancies to the dislocation), thereby avoiding an obstacle. From W. F. Hosford, *Mechanical Behavior of Materials*, 2nd ed., Cambridge University Press (2011).

the rate of climb does depend very strongly on the stress,

$$\dot{\varepsilon} = \mathrm{A_s}\sigma^m. \qquad 12.24$$

The value of m is approximately five for climb-controlled creep[1]. Because climb depends on diffusion, the constant A_s has the same temperature dependence as lattice diffusion. At lower temperatures, creep is not entirely climb-controlled and higher exponents are observed.

Equations 12.21 through 12.24 predict creep rates that depend only on stress and temperature and not on strain. Thus, they apply only to stage II or steady-state creep.

For diffusion-controlled creep,

$$\dot{\varepsilon} = A\sigma^m \exp(-\Delta h/RT) \qquad 12.25$$

where Δh is the activation energy for creep (see Figure 12.17).

[1] This notation widely used in the creep literature is the inverse of that earlier in this chapter where we wrote $\sigma \propto \dot{\varepsilon}^{\mathrm{m}}$ or equivalently, $\dot{\varepsilon} \propto \sigma^{1/m}$. The m used in the creep literature is the reciprocal of the m used earlier. The value of $m = 5$ here corresponds to a value of $m = 0.2$ in the notation used earlier.

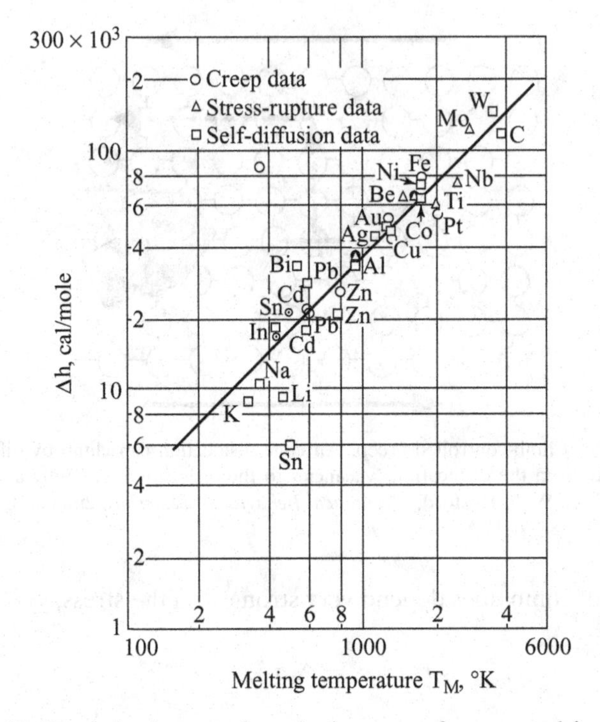

Figure 12.17. Correlation between the activation energy for creep and for diffusion. From E. Dorn. In *Creep and Fracture of Metals at High Temperatures*, Her Majesty's Stationary Office (1956).

DEFORMATION MECHANISM MAPS

The dominant deformation mechanism depends on both temperature and stress level. Figure 12.18 is a deformation map that summarizes the dominant mechanism. Slip is the controlling mechanism at high stresses while diffusion controlled mechanisms dominate at low stresses and high temperatures. Grain boundary diffusion (Coble creep) is most important with small grain sizes.

Multiple mechanisms: More than one creep mechanism may be operating. There are two possibilities; either the mechanisms operate independently or they act cooperatively. If they operate independently the overall creep rate, $\dot{\varepsilon}$, is the sum of the rates due to each mechanism,

$$\dot{\varepsilon} = \dot{\varepsilon}_A + \dot{\varepsilon}_B + \cdots \qquad 12.26$$

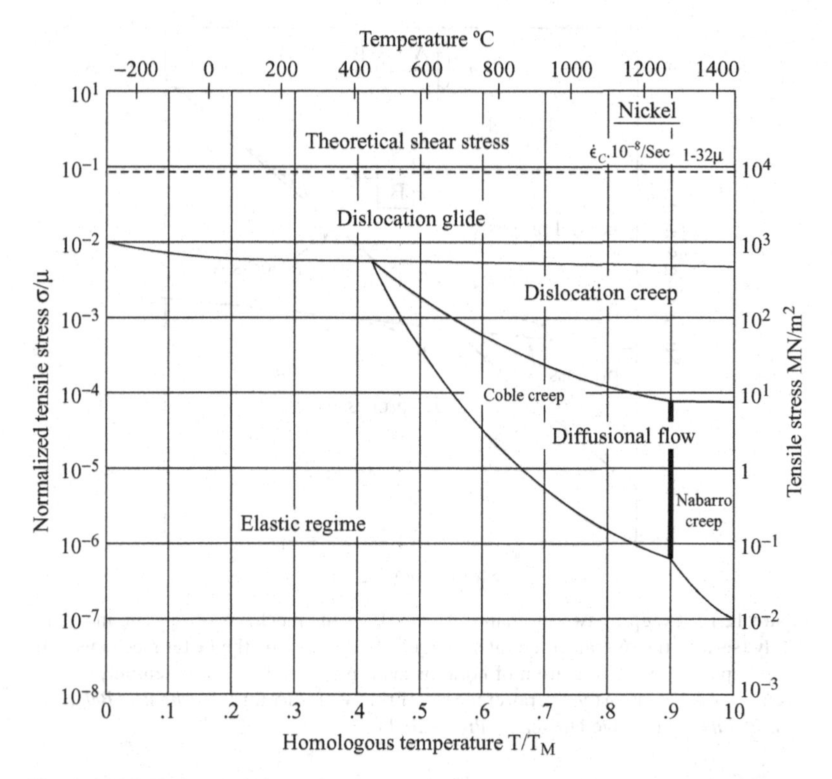

Figure 12.18. Deformation mechanism map for pure nickel with a grain size of $d = 32\ \mu$m. The strain rate is a function of stress and temperature. Different mechanisms are dominant in different regimes. Coble creep is controlled by grain boundary diffusion and Nabarro creep by lattice diffusion. From M. F. Ashby, *Acta Met.* v. 20 (1972).

The result is like two mechanisms acting in series. The overall strain rate depends primarily on the most rapid mechanism.

On the other hand, two mechanisms may be required to operate simultaneously, as in the case grain-boundary sliding requiring another mechanism. Where two parallel mechanisms are required, both must operate at the same rate, so the overall rate is determined by the potentially slower mechanism.

$$\dot{\varepsilon} = \dot{\varepsilon}_A = \dot{\varepsilon}_B. \qquad 12.27$$

These two possibilities, equations 12.26 and 12.27, are illustrated in Figure 12.19.

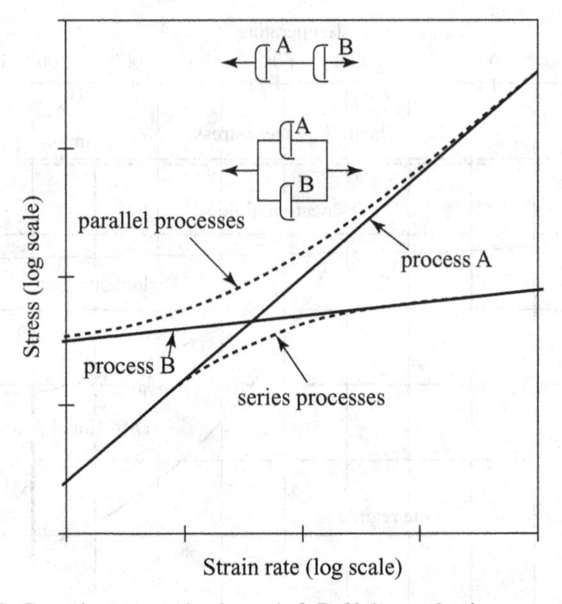

Figure 12.19. Creep by two mechanisms, A & B. If the mechanisms operate independently (series), the overall creep rate is largely determined by the faster mechanism. If creep depends on the operation of both mechanisms (parallel), the potentially slower mechanism will control the overall creep rate. From W. F. Hosford, *Mechanical Behavior of Materials*, Cambridge University Press (2011).

TEMPERATURE DEPENDENCE OF CREEP

Creep can be regarded as a rate process that depends on thermal activation. The simple approach taken by Sherby and Dorn was to assume that for any mechanism

$$\dot{\varepsilon} = f(\sigma)\exp(-Q/RT), \qquad 12.28$$

so that the stress-dependence is incorporated into the pre-exponential term. This is equivalent to the Zener-Hollomon approach. The value of Q depends on whether creep is controlled by lattice diffusion or by grain boundary diffusion.

Sometimes the temperature-dependence of creep rates are expressed in terms of pre-exponential term that is inversely proportional to the temperature, T, for example

$$\dot{\varepsilon} = (A/T)\exp(-Q/RT)[f(\sigma, d)]. \qquad 12.29$$

However, the T in the pre-exponential term does not have a great affect on the temperature dependence.

NOTE OF INTEREST

John Herbert Hollomon, Jr., generally known as Herbert Hollomon, was a prominent American engineer and founding member of the National Academy of Engineering. He was born in Norfolk, Virginia on March 12, 1919. In 1946, he received his ScD from the Massachusetts Institute of Technology (MIT) in metallurgy. He then joined the General Electric company where he eventually became general manager of their laboratories in Schenectady, New York.

In 1962, he was appointed first assistant secretary for science and technology at the United States Department of Commerce where he established the Environmental Sciences Services Administration (later, renamed the National Oceanic and Atmospheric Administration). He served for part of 1967 as acting under secretary of commerce, but left government for the University of Oklahoma where he served as President.

In 1970, Hollomon returned to MIT as consultant to the president and subsequently as Professor of Engineering. In 1983, he moved to the Boston University, where he remained until his death on May 8, 1985.

John Dorn was born April 10, 1909 in Chicago, Illinois. He received his BS (1931) and MS (1932) degrees from Northwestern University in Chemistry and his PhD (1936) in Physical Chemistry at the University of Minnesota. He spent the next two years at Battelle Memorial Institute in Columbus, Ohio. John Dorn then became a faculty member at the University of California at Berkeley where he spent the rest of his career in the field of physical metallurgy. At Berkeley, he was known as an outstanding teacher as well as a superb research scientist. Although trained in chemistry, he set upon the path from chemistry to metallurgy well traveled by many outstanding metallurgists of the 1920s, 1930s, and 1940s. Dorn was particularly famous for his work on the high temperature creep of metals. He and his best-known student, Oleg Sherby, established that the activation energy of high temperature creep and the activation energy of self diffusion are the same.

He authored or coauthored 180 research papers. Suffering from a heart condition and lung cancer, John Dorn died on September 24, 1971.

REFERENCES

1. J. H. Hollomon and L. Jaffe, *Ferrous Metallurgical Design*, Wiley & Sons (1947).
2. E. Voce, *J. Inst. Metals* v. 74 (1948).
3. F. R. Nabarro, *Phys. Soc. London* (1948)
4. J. Weertman, *J. Appl. Phys*. v. 28 (1957).

GENERAL REFERENCES

F. A. McClintock and A. S. Argon, *Mechanical Behavior of Materials*, Addison-Wesley (1966).

John Dorn, *Mechanical Behavior of Materials at elevated Temperatures*, McGraw-Hill (1961).

D. S. Fields and W. A. Backofen, *Trans. ASM*, v. 51 (1959).

O. D. Sherby, J. L. Lytton, and J. E. Dorn, *AIME Trans*. v. 212 (1958).

Z. S. Basinski, *Acta Met*., v. 5 (1957).

13

DEFECT ANALYSIS

LOCALIZATION OF STRAIN AT DEFECTS

If the stresses that cause deformation in a body are not uniform, differences develop between the strains in the different regions. These differences depend on the level of n. If n is high, the difference will be less than if n is low. A tensile bar in which the cross-sectional area of one region, A_{ao}, is a little less than the area, A_{bo}, in the rest of the bar is illustrated in Figure 13.1.

The tensile force, F, is the same in both, $F_a = F_b$, regions so

$$\sigma_a A_a = \sigma_b A_b \qquad 13.1$$

Because $\varepsilon = \ln(A_o/A)$, the instantaneous area, A, may be expressed as $A = A_o \exp(-\varepsilon)$, so $A_a = A_{ao} \exp(-\varepsilon_a)$, $A_b = A_{bo} \exp(-\varepsilon_b)$. Assuming power-law hardening $\sigma_a = K\varepsilon_a^n$, and $\sigma_b = K\varepsilon_b^n$. Substituting into equation 13.1.

$$K\varepsilon_a^n A_{ao} \exp(-\varepsilon_a) = K\varepsilon_b^n A_{bo} \exp(-\varepsilon_b). \qquad 13.2$$

Now simplify by substituting, $f = A_{ao}/A_{bo}$,

$$\varepsilon_b^n \exp(-\varepsilon_b) = f\varepsilon_a^n \exp(-\varepsilon_a). \qquad 13.3$$

To find ε_b as a function of ε_a, equation 13.3 must be evaluated numerically. Figure 13.2 shows that with low values of n, the region with the

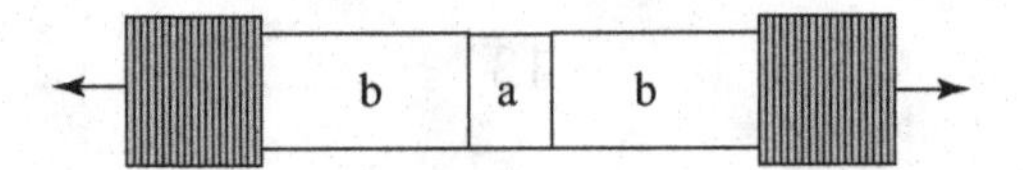

Figure 13.1. The initial cross-sectional area of region a is *f* times the cross sectional area of b.

larger cross section deforms very little. On the other hand, if n is high, there is appreciable deformation in the thicker region so more overall stretching will have occurred when the thinner region fails. This leads to greater formability.

EFFECT OF STRAIN-RATE ON LOCALIZATION

To analyze the effect of strain-rate sensitivity, reconsider the tension test on the stepped bar in Figure 13.1, neglecting strain hardening and assuming that $\sigma = C\dot{\varepsilon}^m$. The bar is divided into two regions, one with an initial cross section of A_{bo} and the other $A_{ao} = fA_{bo}$. As before, the forces must balance, so $\sigma_b A_b = \sigma_a A_a$. Substituting $A_i = A_{io}\exp(\varepsilon_i)$

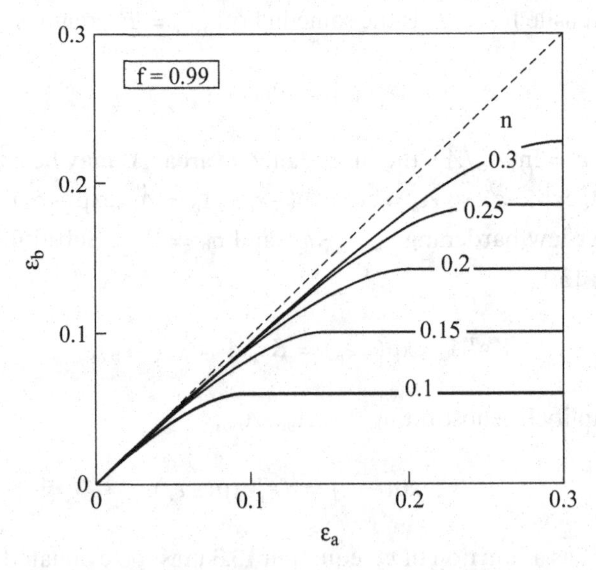

Figure 13.2. The relative strains in two regions of a tensile bar having different initial cross-sectional areas. From W. F. Hosford, *Mechanical Behavior of Materials* 2nd ed., Cambridge University Press 2011.

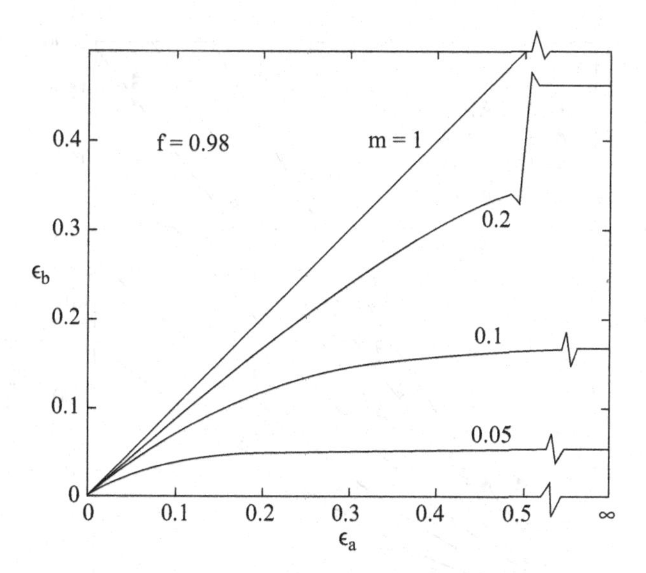

Figure 13.3. Relative strains in unreduced, ε_b, and reduced section, ε_a, of a stepped tensile specimen for various levels of m, assuming no strain hardening. From W. F. Hosford and R. M. Caddell, *Metal Forming: Mechanics and Metallurgy*, 4th ed., Cambridge University Press (2011).

and

$\sigma_i = C\dot{\varepsilon}_i^m$, the force balance becomes

$$A_{bo}\exp(-\varepsilon_b)\dot{\varepsilon}_b^m = A_{ao}\exp(-\varepsilon_a)\dot{\varepsilon}_a^m, \qquad 13.4$$

where $\dot{\varepsilon}_a$ and $\dot{\varepsilon}_b$ are the strain rates in the reduced and unreduced sections. Expressing $\dot{\varepsilon}$ as $d\varepsilon/dt$,

$$\exp(-\varepsilon_b)\left(\frac{d\varepsilon_b}{dt}\right)^m = f\exp(-\varepsilon_a)\left(\frac{d\varepsilon_a}{dt}\right)^m. \qquad 13.5$$

Raising both sides to the $(1/m)$ power and canceling dt

$$\int_o^{\varepsilon_b}\exp(-\varepsilon_b/m)d\varepsilon_b = \int_o^{\varepsilon_a} f^{1/m}\exp(-\varepsilon_a/m)d\varepsilon_a. \qquad 13.6$$

Integration gives

$$\exp(-\varepsilon_b/m) - 1 = f^{1/m}[\exp(-\varepsilon_a/m) - 1]. \qquad 13.7$$

Numerical solutions of ε_b as a function of ε_a for $f = 0.98$ and several levels of m are shown in Figure 13.3. At low levels of m (or low values

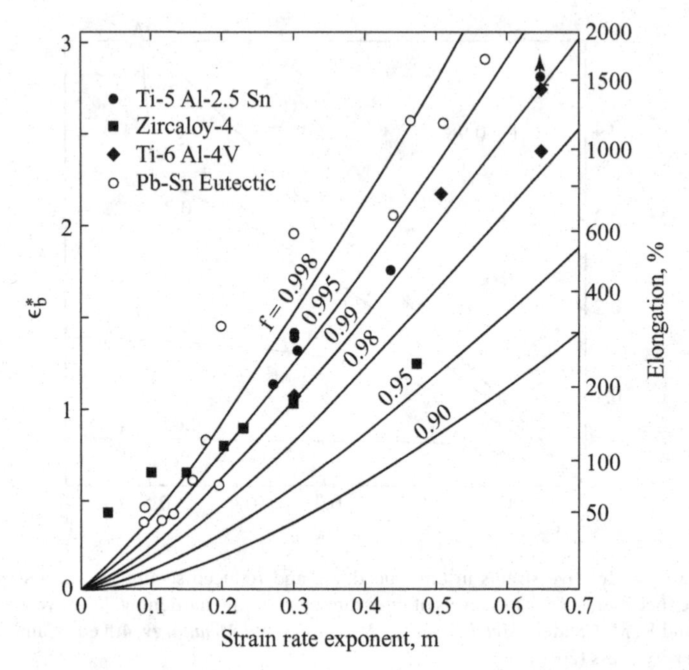

Figure 13.4. Limiting strains, ε_b^*, in unreduced sections of stepped tensile specimens as a function of m and f. Values of percent elongation corresponding to ε_b^* are indicated on the right. From W. F. Hosford and R. M. Caddell, *Metal Forming: Mechanics and Metallurgy*, 4th ed., Cambridge University Press (2011).

of f), ε_b tends to saturate early and approaches a limiting strain ε_b^* at moderate levels of ε_a, but with higher m-values, saturation of ε_b is much delayed, that is, localization of strain in the reduced section (or the onset of a sharp neck) is postponed. Thus, the conditions that promote high m-values also promote high failure strains. Letting $\varepsilon_a \to \infty$ in equation 13.7 will not cause great error and will provide limiting values for ε_b^*. With this condition,

$$\varepsilon_b^* = -m \ln(1 - f^{1/m}). \tag{13.8}$$

In Figure 13.4, values of ε_b^* calculated from equation 13.8 are plotted against m for various levels of f. The values of tensile elongation corresponding to ε_b^* are indicated on the right margin. It is now clear why large elongations are observed under superplastic conditions. The data suggest an inhomogeneity factor of about 0.99 to 0.998 (for a round bar 0.250-in. diameter, a diameter variation of 0.0005 in. corresponds to $f = 0.996$). The general agreement is perhaps fortuitous considering the assumptions and simplifications. The values of ε_b^* are the strains away from the neck, so the total elongation would be even higher than indicated here. On the other hand, strains in the neck, ε_a, are not infinite, so the ε_b^* values corresponding to realistic maximum values of ε_a will be lower than those used in Figure 13.4. Also, the experimental values are affected by difficulties in maintaining constant temperature over the length of the bar as well as a constant strain rate in the deforming section. Nevertheless, the agreement between theory and experiments is striking.

COMBINED EFFECTS OF STRAIN HARDENING AND STRAIN-RATE SENSITIVITY

The combined effects of strain hardening and strain-rate sensitivity on strain localization can be analyzed by reconsidering a tension test on an inhomogeneous specimen with two regions of initial cross-sectional areas A_{bo} and $A_{ao} = fA_{bo}$. Now assume that the material strain hardens and is also rate sensitive, so that the flow stress is given by

$$\sigma = C'\varepsilon^n\dot{\varepsilon}^m. \qquad 13.9$$

Substituting $A_i = A_{io}\exp(-\varepsilon_i)$ and $\sigma = C'\varepsilon^n\dot{\varepsilon}^m$ into a force balance, $A_b\,\sigma_b = A_a\,\sigma_a$, results in $A_{bo}\exp(-\varepsilon_b)\varepsilon_b^n\dot{\varepsilon}_b^m = A_{ao}\exp(-\varepsilon_a)\varepsilon_a^n\dot{\varepsilon}_a^m$. Following the procedure that produced equation 13.7,

$$\int_o^{\varepsilon_b} \exp(-\varepsilon_b/m)\varepsilon_b^{n/m}d\varepsilon_b = f^{1/m}\int_o^{\varepsilon_a} \exp(-\varepsilon_a/m)\varepsilon_a^{n/m}d\varepsilon_a. \qquad 13.10$$

The results of integration and numerical evaluation are shown in Figure 13.5, where ε_b is plotted as a function of ε_a for $n = 0.2$, $f = 0.98$, and several levels of m. It is apparent that even quite low levels

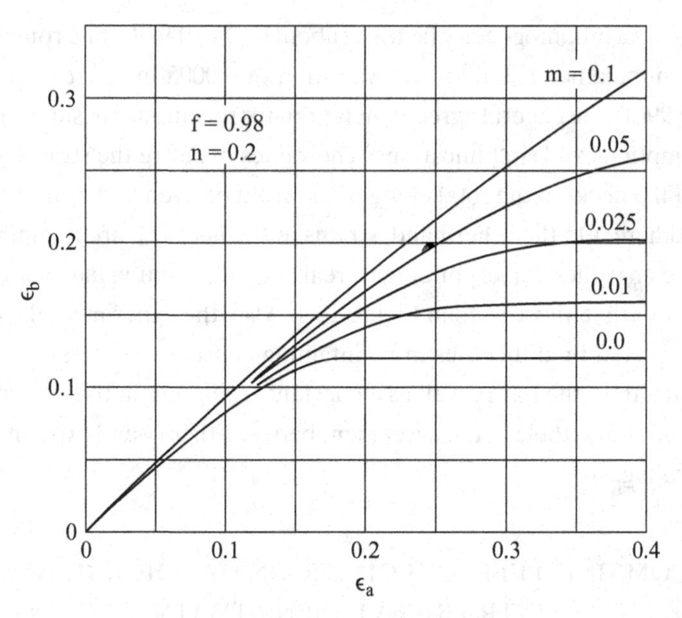

Figure 13.5. Comparison of trains in reduced an unreduced regions of a tensile bar calculated with $f = 0.98$ and $n = 0.20$. Note that even relatively low levels of m influence È$_b$. From W. F. Hosford and R. M. Caddell, *Metal Forming: Mechanics and Metallurgy*, 4th ed., Cambridge University Press (2011).

of m play a significant role in controlling the strains reached in the unnecked region of the bar.

LOCALIZED NECKING

Many sheet-forming operations involve biaxial stretching in the plane of the sheet. Failures occur by the formation of sharp local necks. Localized necking should not be confused with diffuse necking, which precedes failure in round tensile specimens. Diffuse necking of sheet specimens involves contraction in both the lateral and width directions. In sheet tensile specimens, local necking occurs after diffuse necking. During local necking, the specimen thins without further width contraction. Figure 13.6 illustrates local necking. At first, the specimen elongates uniformly. At maximum load, a diffuse neck forms by

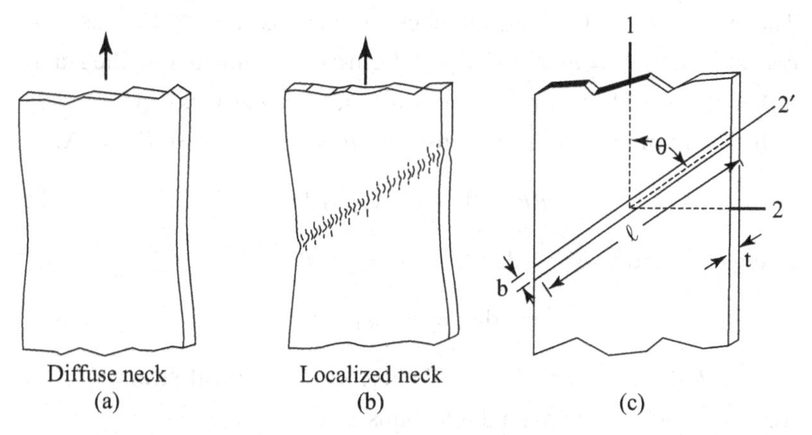

Figure 13.6. Diffuse neck (a) localized neck (b) coordinate system used in analysis (c). From W. F. Hosford and R. M. Caddell, *Metal Forming: Mechanics and Metallurgy*, 4th ed., Cambridge University Press (2011).

contraction of both the width and thickness when $\varepsilon_1 = n$. (Figure 13.6a). Finally a local neck develops (Figure 13.6b).

In a tension test the strain in the width direction cannot localize easily, but eventually a condition is reached where a sharp local neck can form at some characteristic angle, θ, to the tensile axis. Typically the width of the neck is roughly equal to the thickness so very little elongation occurs after local necking. The strain parallel to the neck, $d\varepsilon_{2'}$, must be zero, but

$$d\varepsilon_{2'} = d\varepsilon_1 \cos^2\theta + d\varepsilon_2 \sin^2\theta = 0. \qquad 13.11$$

For an isotropic material under uniaxial tension in the 1-direction, $d\varepsilon_2 = d\varepsilon_3 = -d\varepsilon_1/2$. Substituting into equation 13.11, $\cos^2\theta - \sin^2\theta/2 = 0$, or

$$\tan\theta = \sqrt{2},\ \theta = 54.74° \qquad 13.12$$

If the metal is anisotropic, $d\varepsilon_2 = -R/(R+1)d\varepsilon_1$. 13.13

In this case,

$$\tan\theta = \sqrt{(R+1)/R}. \qquad 13.14$$

The cross-sectional area of the neck, A', itself is $A' = \ell t$. Because ℓ is constant, $dA'/A' = dt/t = d\varepsilon_3$, and the area perpendicular to the 1-axis is $A = A' \sin\theta$. However, θ is also constant so a local neck can form only if the load can fall under the constraint $d\varepsilon_{2'} = 0$. Because $F = \sigma_1 A$,

$$dF = 0 = \sigma_1 dA + A d\sigma_1. \qquad 13.15$$

or $d\sigma_1/\sigma_1 = -dA/A = -d\varepsilon_3$. Since $d\varepsilon_3 = -d\varepsilon_1/2$,

$$d\sigma_1/\sigma_1 = d\varepsilon_1/2. \qquad 13.16$$

If $\sigma_1 = K\varepsilon_1^n$, $d\sigma_1 = nK\varepsilon_1^{n-1}d\varepsilon_1$. Therefore the critical strain, ε^*, for localized necking in uniaxial tension is

$$\varepsilon_1^* = 2n. \qquad 13.17$$

In sheet forming, the stress state is rarely uniaxial tension but the same principles can be used to develop the conditions for localized necking under a general stress state of biaxial tension. Assume that the strain ratio $\rho = \varepsilon_2/\varepsilon_1$ remains constant during loading. (This is equivalent to assuming that the stress ratio $\alpha = \sigma_2/\sigma_1$ remains constant.) Substitution of $\varepsilon_2/\varepsilon_1 = \rho$ into equation 13.11 gives $\varepsilon_1\cos^2\theta + \rho\varepsilon_2\sin^2\theta = 0$, or

$$\tan\theta = 1/\sqrt{-\rho}. \qquad 13.18$$

The angle, θ, can have a real value only if ρ is negative (that is, ε_2 is negative). If ρ is positive, there is no angle at which a local neck can form.

The critical strain for localized necking is also influenced by ρ. For constant volume, $d\varepsilon_3 = -(1 + \rho)d\varepsilon_1$. Substituting into $d\sigma_1/\sigma_1 = -d\varepsilon_3$,

$$d\sigma_1/\sigma_1 = (1 + \rho)d\varepsilon_1. \qquad 13.19$$

With power-law, hardening the condition for local necking is

$$\varepsilon_1^* = \frac{n}{1+\rho}. \qquad 13.20$$

Equation 13.20 implies that the critical strain for localized necking, ε_1^*, decreases from $2n$ in uniaxial tension to zero for plane-strain tension.

The previous analysis seems to imply that local necks cannot form if ε_2 is positive. However, this is true only if ρ and α remain constant

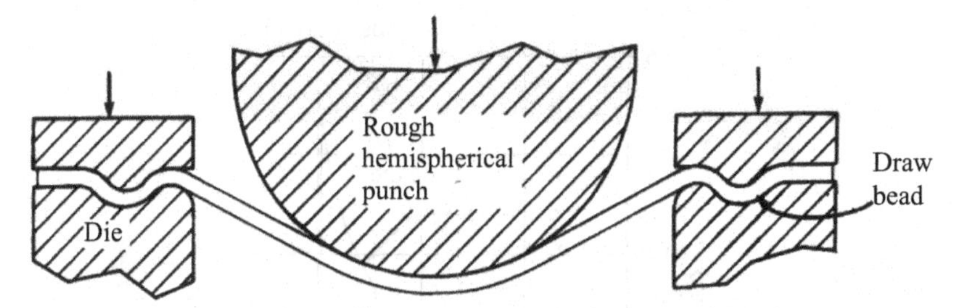

Figure 13.7. Sketch of a rough hemispherical punch. From W. F. Hosford and R. M. Caddell, *Metal Forming: Mechanics and Metallurgy*, 4th ed., Cambridge University Press (2011).

during loading. If they do change during stretching, local necking can occur even with ε_2 being positive. What is critical is that $\rho' = d\varepsilon_2/d\varepsilon_1$ become zero rather ratio, $\rho = \varepsilon_2/\varepsilon_1$, of total strains be zero.

Often, tool geometry causes a change of strain path. Consider a sheet being stretched over a hemispherical dome as shown in Figure 13.7. The flange is locked to prevent drawing. If friction between punch and sheet is high enough to prevent sliding, deformation ceases where the sheet contacts the punch. Elements in the region between the punch and die are free to expand biaxially. However, as the element approached the punch the circumferential strain is constrained by neighboring material on the punch so $d\varepsilon_2/d\varepsilon_1 \to 0$.

It has been argued [3, 4] that because of variations of sheet thickness, grain size, texture, or solute concentration, there can exist local troughs that are softer than the surroundings that lie perpendicular to the 1-axis (Figure 13.8). Although such a trough is not a true neck, it can develop into one. The strain, ε_1, in the trough will grow faster than ε_1 outside of it, but the strain, ε_2, in the trough can grow only at the same as outside the trough. Therefore, the local value of $\rho' = d\varepsilon_2/d\varepsilon_1$

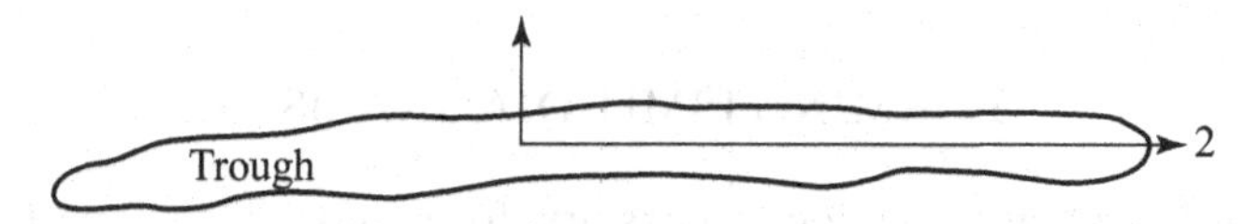

Figure 13.8. Sketch of a trough parallel to the 2-axis. From W. F. Hosford and R. M. Caddell, *Metal Forming: Mechanics and Metallurgy*, 4th ed., Cambridge University Press (2011).

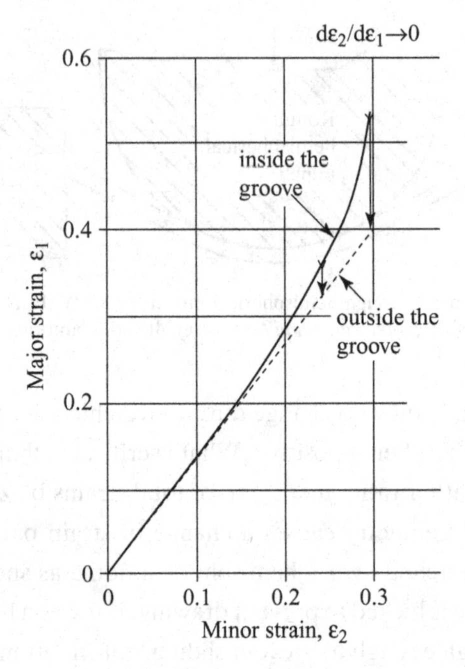

Figure 13.9. Strain paths inside and outside a preexisting groove for a linear strain path imposed outside the groove. The forming limit corresponds to the strain outside the groove when $d\varepsilon_2/d\varepsilon_1 \to \infty$. From Alejandro Graf and W.F. Hosford in *Forming Limit Diagrams, Concepts, Methods and Applications*, TMS (1989).

decreases during stretching. Once ρ' reaches zero, a local neck can form. The trough consists of material that is either thinner or weaker than material outside of it.

As the strain rate, $\dot{\varepsilon}_1$, inside the groove accelerates, the ratio, $\dot{\varepsilon}_2/\dot{\varepsilon}_1$, approaches zero. Figure 13.9 shows how the strain paths inside and outside of the groove can diverge. The terminal strain outside the groove is the limit strain. Very shallow grooves are sufficient to cause such localization. How rapidly this happens depends on n and to a lesser extent on m.

FORMING LIMIT DIAGRAMS

The strains, ε_1^*, at which local necks first form, have been experimentally observed for a wide range of sheet materials and loading paths. A plot of the combination of strains that lead to failure is called a

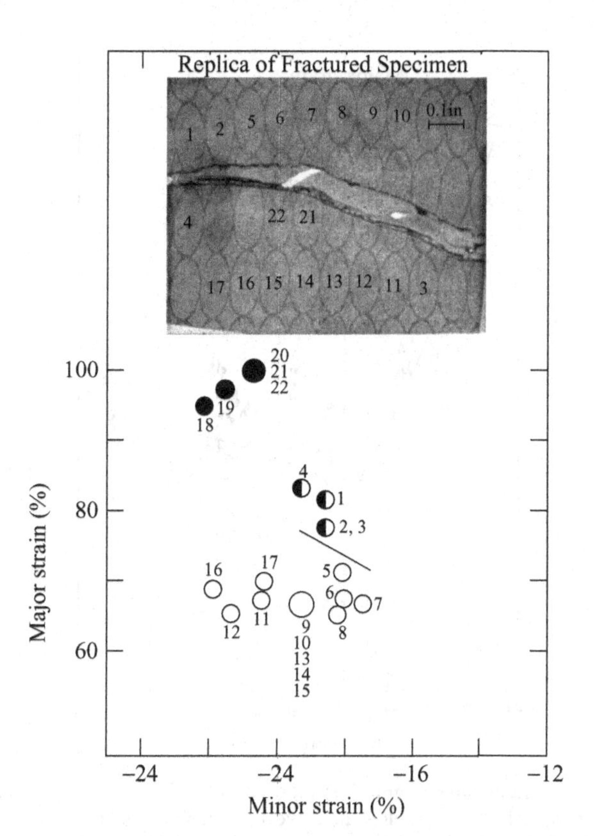

Figure 13.10. Distortion of printed circles near a localized neck and a plot of the strains in the circles. Solid points are for grid circles through which the failure occurred. Open points are for grid circles well removed from the failure and partially filled circles are for grid circles very near the failure. From S. S. Hecker, *Sheet Metal Industries*, v. 52 (1975).

forming limit diagram or FLD. Combinations of $\dot{\varepsilon}_1$ and $\dot{\varepsilon}_2$ below the curve are safe and those above the curve lead to failure. Note that the lowest value of $\dot{\varepsilon}_1$ corresponds to plane strain, $\varepsilon_1 = 0$. A widely used technique is to print or etch a grid of small circles of diameter, d_o, on the sheet before deformation. As the sheet is deformed, the circles become ellipses. The principal strains can be found by measuring the major and minor diameters after straining. By convention, the engineering strains e_1 and e_2 are reported. These values at the neck or fracture give the failure condition, while the strains away from the failure indicate *safe* conditions as shown in Figure 13.10. A plot of

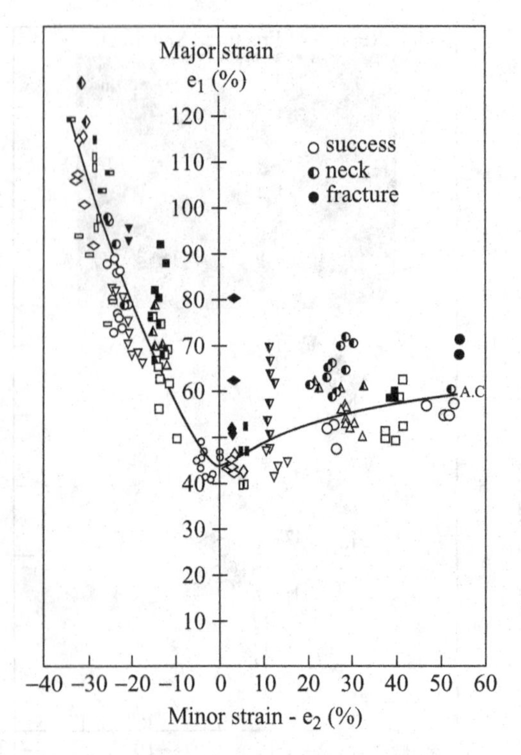

Figure 13.11. Forming limit diagram for a low-carbon steel determined from data like that in Figure 16.6. The strains below the curve are acceptable but those above the curve correspond to failure. From S. S. Hecker, *Sheet Metal Industries*, v. 52 (1975).

these measured strains for a material form its forming limit diagram (FLD) also called a Keeler-Goodwin diagram, because of early work by Keeler [5] and Goodwin [6]. Figure 13.11 and theoretical curves are parallel. The fact that the experimental curve is higher reflects the fact that a neck has to be developed before it can be detected.

CALCULATION OF FORMING LIMIT DIAGRAMS

The left-hand side of the forming limit diagrams correspond to $\varepsilon_1 = n/(1+\rho)$, but because equation 13.18 has no real solution for $\rho < 1$, the right-hand side of the forming limit diagram appeared to be impossible. Marciniak and Kuczynski [3, 4], however, showed that the right-hand side of the forming limit diagram for a material can be

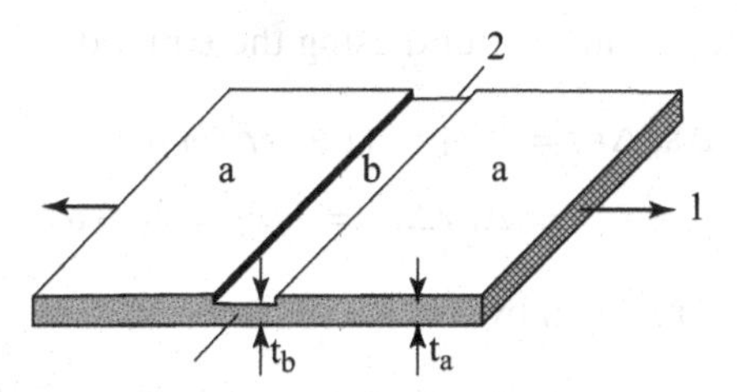

Figure 13.12. Schematic illustration of a pre-existing defect in a sheet. From Alejandro Graf and W. F. Hosford. In *Forming Limit Diagrams, Concepts, Methods and Applications*, TMS (1989).

calculated by assuming that there is a pre-existing defect, which lies perpendicular to the major stress axis. For calculation purposes, this defect can be approximated as a region that is thinner than the rest of the sheet. Figure 13.12 illustrates this sort of defect. A ratio of the initial thicknesses inside and outside the defect, $f = t_{bo}/t_{ao}$ is assumed. Also, it is assumed that the stress ratio, α_a, outside of the groove remains constant during loading.

The calculations are based on imposing strain increments $\Delta\varepsilon_{b1}$ on the material inside of the defect and finding the corresponding value of $\Delta\varepsilon_a$. This involves an iterative procedure. First, a value of $\Delta\varepsilon_a$ must be guessed. (It will be somewhat less than $\Delta\varepsilon_{b1}$.) This value is used to calculate $\Delta\varepsilon_{a2} = \rho_a\Delta\varepsilon_{a1}$. Compatibility requires that $\Delta\varepsilon_{b2} = \Delta\varepsilon_{a2}$.

Then,

$$\rho_b = \Delta\varepsilon_{b2}/\Delta\varepsilon_{b1} = \dot{\varepsilon}_{2b}/\dot{\varepsilon}_{1b}, \quad \rho_a = \Delta\varepsilon_{a2}/\Delta\varepsilon_{a1} = \dot{\varepsilon}_{2a}/\dot{\varepsilon}_{1a}. \qquad 13.21$$

Next, α_a and α_b can be found from the associated flow rule. For Hill's criterion,

$$\alpha = [(R+1)\rho + R]/[(R+1) + R\rho]. \qquad 13.22a$$

However, with the high exponent yield criterion, the flow rule

$$\rho = [\alpha^{a-1} - R(1-\alpha)^{a-1}]/[1 + R(1-\alpha)^{a-1}] \qquad 13.22b$$

must be solved by iteration.

Then, β_b and β_a can be found using the equation

$$\beta = \Delta\bar{\varepsilon}/\Delta\varepsilon_1 = \dot{\bar{\varepsilon}}/\dot{\varepsilon}_1 = (1+\alpha\rho)/\varphi\beta$$
$$= \Delta\bar{\varepsilon}/\Delta\varepsilon_1 = \dot{\bar{\varepsilon}}/\dot{\varepsilon}_1 = (1+\alpha\rho)/\varphi, \qquad 13.23$$

where φ_b and φ_a are given by

$$\varphi = \bar{\sigma}/\sigma_1 = \{[\alpha^a + 1 + (1-\alpha)^a]/(R+1)\}^{1/a}. \qquad 13.24$$

The thickness strain rate, $\dot{\varepsilon}_3 = \ln \dot{t}/t$, is also given by

$$\ln \dot{t}/t = -(1+\rho)\dot{\varepsilon} = -(1+\rho)\dot{\varepsilon}_2/\rho. \qquad 13.25$$

The effective stress is given by the effective stress-strain relation

$$\bar{\sigma} = K\dot{\bar{\varepsilon}}^m \bar{\varepsilon}^n. \qquad 13.26$$

In the calculations, the value of $\bar{\varepsilon}$ is incremented by $\Delta\bar{\varepsilon}$ so $\bar{\sigma}$ is calculated as

$$\bar{\sigma} = K\dot{\bar{\varepsilon}}^m(\bar{\varepsilon} + \Delta\bar{\varepsilon})^n. \qquad 13.27$$

Letting F_1 be the force per length normal to the groove

$$\sigma = \varphi F_1/t, \qquad 13.28$$

and $\sigma = \varphi F_1/hF_1$,

$$\dot{\bar{\varepsilon}} = \beta\dot{\varepsilon}_1 = -\beta(\dot{t}/t)/(\rho+1). \qquad 13.29$$

Combining equations 13.26, 13.27, 13.28, and 13.29 results in

$$F_1 = K(h/\varphi)(\bar{\varepsilon} + \Delta\bar{\varepsilon})^n(\beta\varepsilon_2/\rho)^m. \qquad 13.30$$

The values of F_1 and $\dot{\varepsilon}_2$ must be the same inside and outside of the groove, so

$$(t_a/\varphi_a)(\bar{\varepsilon}_a + \Delta\bar{\varepsilon}_a)^n(\beta_a\varepsilon_{2a}/\rho_a)^m = (t_b/\varphi_b)(\bar{\varepsilon}_b + \Delta\bar{\varepsilon}_b)^n(\beta_b\varepsilon_{2b}/\rho_b)^m. \qquad 13.31$$

Finally substituting $f = t_{bo}/t_{ao}$ and $t = t_o \exp(\varepsilon_3)$ and $t = t_o \exp(\varepsilon_3)$,

$$(\bar{\varepsilon}_a + \Delta\bar{\varepsilon}_a)^n(\beta_a/\rho_a)^m/\varphi_a = f(\bar{\varepsilon}_b + \Delta\bar{\varepsilon}_b)^n(\beta_b/\rho_b)^m/\varphi_b. \qquad 13.32$$

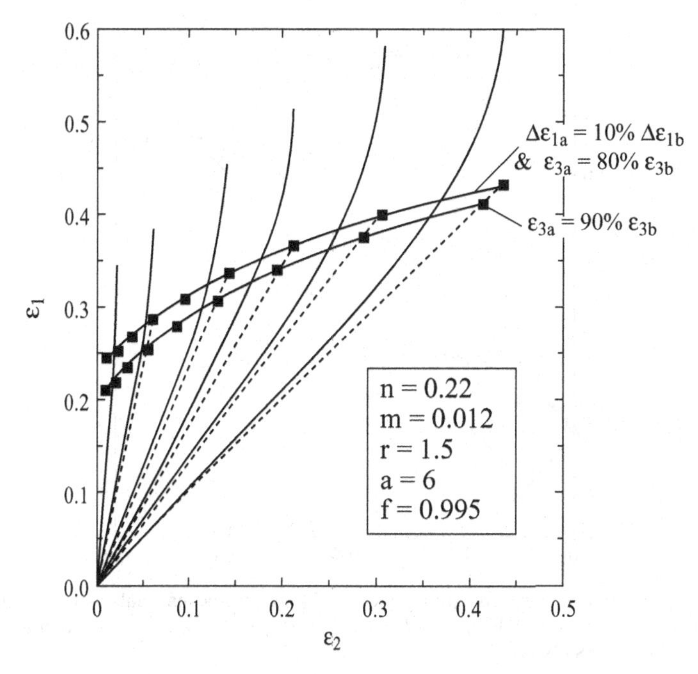

Figure 13.13. Calculated forming limit diagram for a hypothetical material using the high exponent yield criterion. From Alejandro Graf and W. F. Hosford. In *Forming Limit Diagrams, Concepts, Methods and Applications*, TMS (1989).

For each strain increment, $\Delta\varepsilon_{1b}$, in the groove there is a corresponding strain increment,$\Delta\varepsilon_{1a}$, outside the groove. The procedure is to impose a strain increment, $\Delta\bar{\varepsilon}_b$, and then guess the resulting value of $\Delta\varepsilon_{1a}$, and use this value together with α_a to calculate $\beta_b, \varphi_b, \rho_b$, and $\bar{\varepsilon}_b$. These are then substituted into equation 13.32 to find $\Delta\bar{\varepsilon}_a$ and then calculate $\Delta\varepsilon_{1a}$ from the change in $\Delta\bar{\varepsilon}_a$. This value of $\Delta\varepsilon_{1a}$ is compared to the assumed value. This process is repeated until the difference between the assumed and calculated values becomes negligible.

Additional strain increments, $\Delta\varepsilon_{b1}$, are imposed until the $\Delta\varepsilon_{a1} < 0.10\Delta\varepsilon_{b1}$ or some other criterion of approaching plane strain is reached. The values of ε_{a2} and ε_{a1} at this point are taken as points on the FLD. Figure 13.13 shows an example of the calculated strain paths

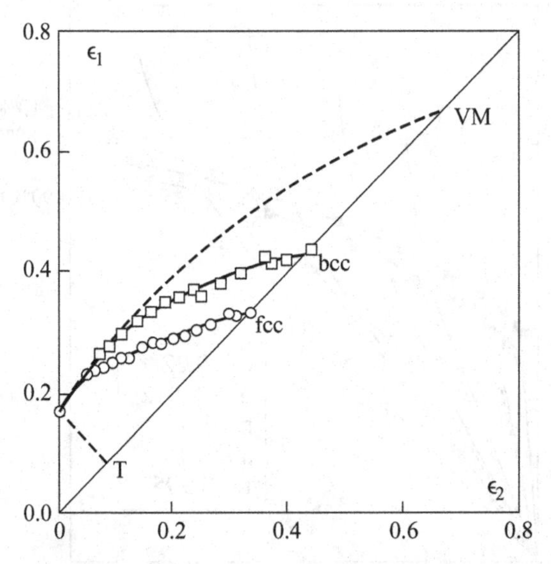

Figure 13.14. Forming limit diagrams calculated for isotropic materials based on different yield criteria. From F. Barlat. In *Forming Limit Diagrams, Concepts, Methods and Applications*, TMS (1989).

inside and outside of the defect using $\Delta\varepsilon_{a1} < 0.10\Delta\varepsilon_{b1}$, $\varepsilon_{3a} = 0.80\varepsilon_{3b}$ and $\varepsilon_{3a} = 0.90\varepsilon_{3b}$ as criteria for failure.

Calculated forming limit diagrams are very sensitive to the yield criterion used in the calculations. Figure 13.14 shows how strongly the assumed yield criterion affects the calculated forming limit diagram.

Figure 13.15 shows that the forming limit diagrams calculated with Hill's 1948 yield criterion are very dependent on the R-value. This is not in accord with experimental observations, which show no appreciable dependence on R. Figure 13.16 shows that with the high exponent criterion, there is virtually no calculated dependence on the R-value.

The reason that the calculated forming limit diagrams are so sensitive to the assumed yield criterion is that for localized necking to occur, the local stress state must change from biaxial stretching

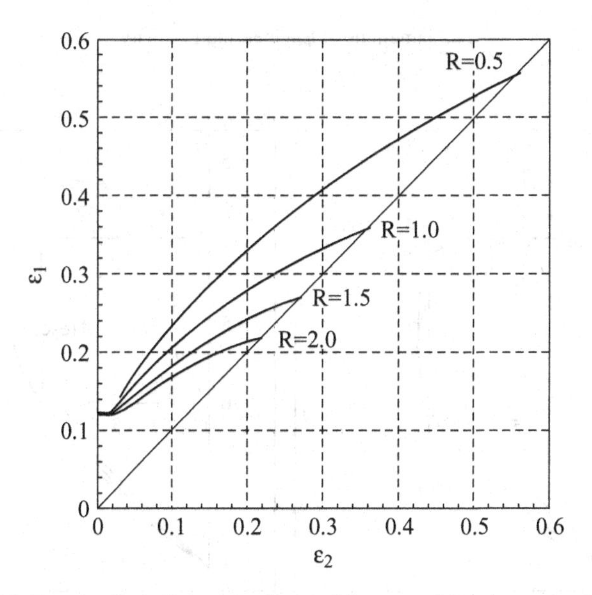

Figure 13.15. Forming limit diagrams calculated for several R-values using Hill's 1948 yield criterion. Values of $n = 0.20$, $m = 0$ and $f = 0.98$ were assumed. From Alejandro Graf and W. F. Hosford. In *Forming Limit Diagrams, Concepts, Methods and Applications*, TMS (1989).

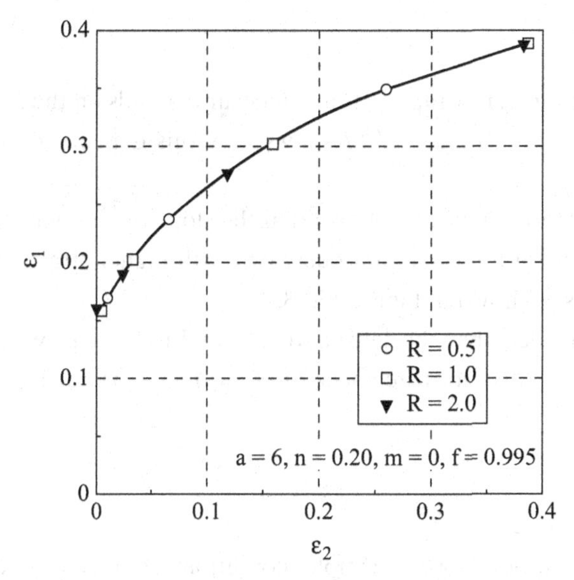

Figure 13.16. Forming limit diagrams calculated for several R-values using $a = 6$ in the high exponent yield criterion. Values of $n = 0.20$, $m = 0$, and $f = 0.995$ were assumed. From Alejandro Graf and W. F. Hosford. In *Forming Limit Diagrams, Concepts, Methods and Applications*, TMS (1989).

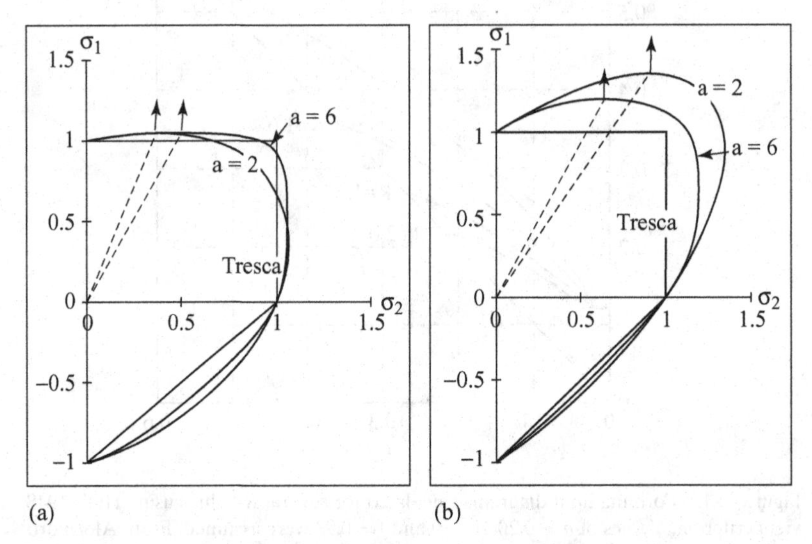

Figure 13.17. Yield criteria for the high exponent criterion and for Hill's '48 criterion for $R = ½$ (left) and $R = 2$ (right). Note that with the high exponent criterion, the stress ratio for plane strain remains near ½ regardless of R. From W. F. Hosford and J. L. Duncan, *JOM* v. 51 (1999).

to plane strain. How rapidly this can occur depends on the stress state for plane strain. Figure 13.17 shows how this depends on the yield criterion.

Calculations have also shown that the slope on the right-hand side of the forming limit diagram decreases with higher strain hardening exponents as shown in Figure 13.18.

Hill showed that the left-hand side of FLD is easily calculated. Equation 13.20 for localized necking, $\varepsilon_1^* = n/(1+\rho)$, can be expressed as

$$\varepsilon_3^* = -n, \tag{13.33}$$

which corresponds to a simple condition of a critical thickness strain.

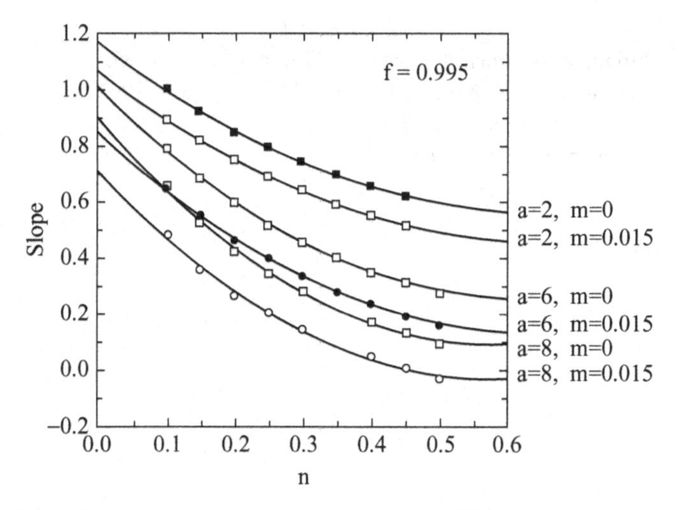

Figure 13.18. The calculated mean slope between $\rho = 0.2$ and 0.8 decreases with the strain hardening exponent, n, and with increasing strain-rate sensitivity, m. From Alejandro Graf and W. F. Hosford. In *Forming Limit Diagrams, Concepts, Methods and Applications*, TMS (1989).

NOTES OF INTEREST

Professor Zdzislaw Marciniak of the Technical University of Warsaw was the first to analyze the effect of a small defect or inhomogeneity on the localization of plastic flow. His analysis of necking forms the basis for understanding superplastic elongations and for calculating forming limit diagrams. He was the Acting Rector, Senior Professor, and Director of the Institute of Metal Forming at the Technical University of Warsaw and has published a number of books in Polish on metal forming and many papers in the international literature.

Stuart Keeler was awarded a S.B. from Ripon College in 1957 and an Sc.D. in Mechanical Metallurgy by MIT in 1961. His doctoral thesis under Professor W. A. Backofen, "Plastic Instability and Fracture in Sheets Stretched Over Rigid Punches," became the foundation for the forming limit diagrams.

REFERENCES

1. Z. Marciniak, *Archiwum Mechanikj Stosowanaj* v. 4, 1965.
2. Z. Marciniak and K. Kuczynski, *Int. J. Mech. Sci.* v. 9 (1967).
3. S. P. Keeler, SAE paper 680092, 1968.
4. G. Goodwin, SAE paper 680093, 1968.

14

EFFECTS OF PRESSURE AND SIGN OF STRESS STATE

The effects of pressure on the yield locus can be confused with the effects of the sign of the stress. For example, twinning is sensitive to the sign of the applied stresses and causes the yield behavior under compression to be different from that under tension.

S-D EFFECT

With a so-called strength differential (SD) effect in high strength steels [1], yield strengths under tension are lower than under compression. The fractional magnitude of the effect, $2(|\sigma_c| - \sigma_T)/[(|\sigma_c| + \sigma_T]$ is between 0.10 and 0.20. Figure 14.1 shows the effect in an AISI 4330 steel and Figure 14.2 indicates that it is a pressure effect.

Although the flow rules, equation 4.18, predict a volume increase with yielding, none has been observed.

POLYMERS

For polymers, the stress-strain curves in compression and tension can be quite different. Figures 14.3 and 14.4 are stress strain curves for epoxy and PMMA in tension and compression. Figure 14.5 compares the yield strengths of polycarbonate as tested in tension, shear and compression. The effect of pressure on the yield strength of PMMA is plotted in Figure 14.6.

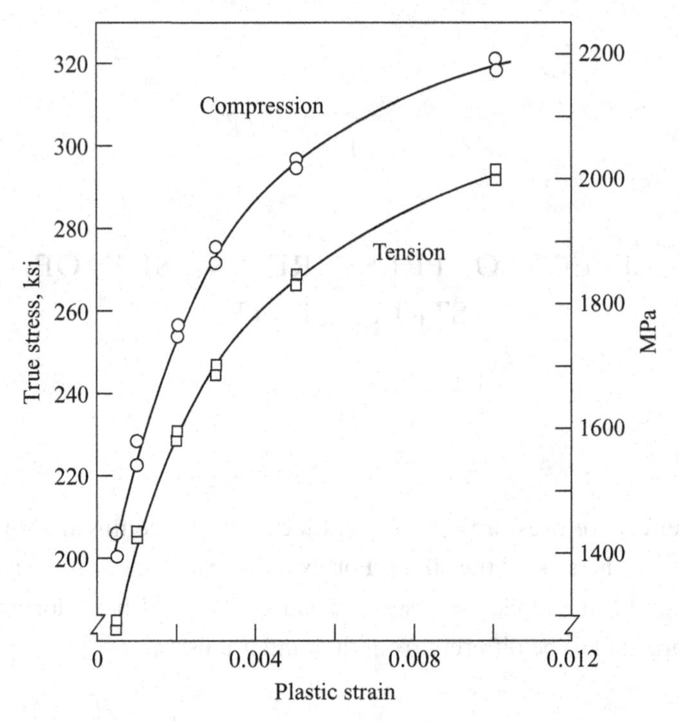

Figure 14.1. True stress strain curves for as-quenched AISI 4330 steel. From G. C. Rauch and W. C. Leslie, *Met. Trans.* v. 3 (1972).

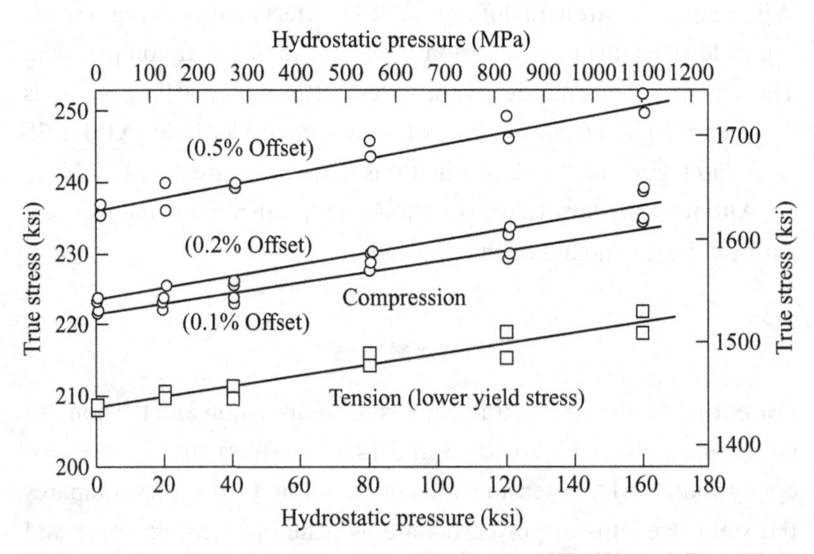

Figure 14.2. The effect of pressure on the SD effect in AISI 4330 steel. From W. A. Spitzig, R. J. Sober, and O. Richmond, *Acta Met.* v. 23 (1975).

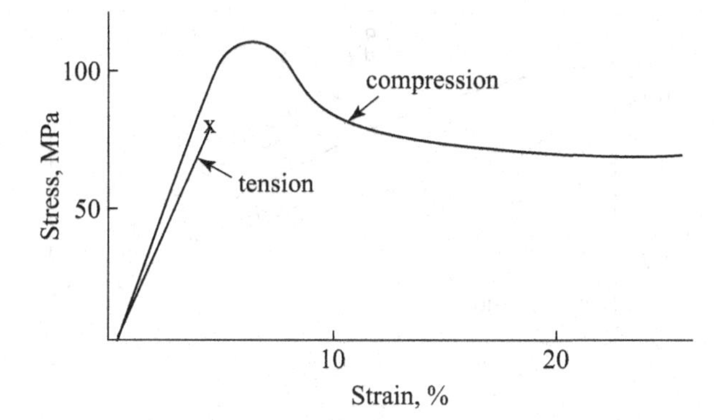

Figure 14.3. Stress strain curves of an epoxy measured in tension and compression tests. From W. F. Hosford, *Mechanical Behavior of Materials*, 2nd ed., Cambridge University Press (2010). Data from ref P. A. Young and R. J. Lovell, *Introduction to Polymers*, 2nd Ed. Chapman and Hall (1991).

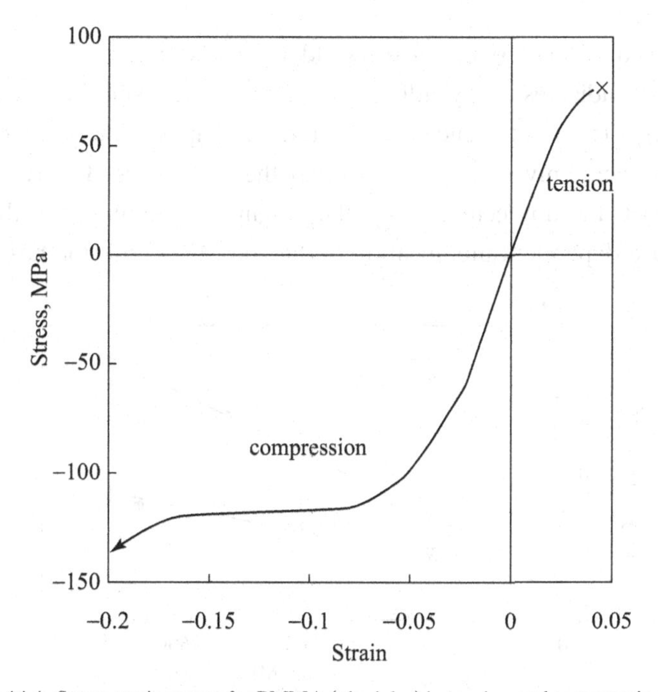

Figure 14.4. Stress strain curves for PMMA (plexiglas) in tension and compression. Note that the strength in compression is higher than that in tension. From W. F. Hosford, *Mechanical Behavior of Materials*, 2nd ed., Cambridge University Press (2010). Data from ref. C. W. Richards, *Engineering Materials Science*, Wadsworth (1961).

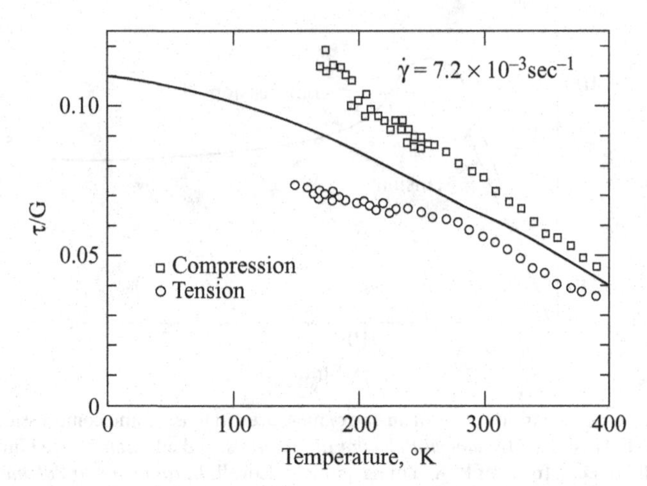

Figure 14.5. Yield strengths of polycarbonate in tension, shear and compression as functions of temperature. From A. S. Argon, *Phil. Mag.* v. 28 (1973).

The difference between the yield strengths in tension and compression indicates that yielding is sensitive to the level of hydrostatic pressure. This dependence is evident in the shapes of the yield loci of several randomly oriented (isotropic) thermoplastics (Figure 14.7). The loci are not centered on the origin. To account for this, a pressure-dependent modification of the von Mises criterion has been

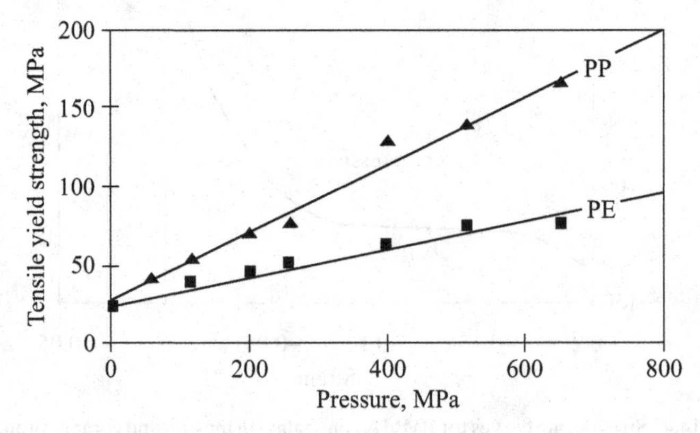

Figure 14.6. The effect of pressure on the yield stresses of polyethylene and polypropylene. From W. F. Hosford, *Mechanical Behavior of Materials*, 2nd ed., Cambridge University Press (2010). Data from Mears, Pae and Sauer, *J. Appl. Phys.* v. 40 (1969).

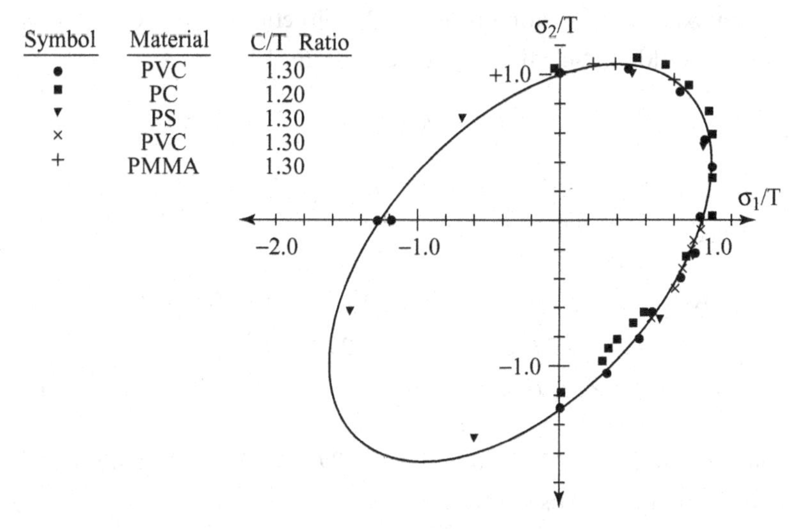

Figure 14.7. Pressure-dependent yield loci. The solid line represents the predictions of equation 14.1, while the dashed line represents the von Mises criterion. From R. J. Ragava, PhD Thesis, University of Michigan (1972).

suggested [2]:

$$(\sigma_2 - \sigma_3)^2 + (\sigma_3 - \sigma_1)^2 + (\sigma_1 - \sigma_2)^2 + 2(C - T)(\sigma_1 + \sigma_2 + \sigma_3) = 2CT, \quad 14.1$$

where T is the yield strength in tension and C is the absolute value of the yield strength in compression. Another possible pressure-dependent modification of the von Mises yield criterion is:

$$(\sigma_2 - \sigma_3)^2 + (\sigma_3 - \sigma_1)^2 + (\sigma_1 - \sigma_2)^2 + \mathrm{K}_1(\sigma_1 + \sigma_2 + \sigma_3)|\sigma_1 + \sigma_2 + \sigma_3| = \mathrm{K}_2. \quad 14.2$$

The constants, K_1 and K_2, in equation 14.1 can be expressed in terms of the tensile yield strength, T, and the absolute magnitude of the compressive yield strength, C. In a uniaxial tensile test in the 1-direction $\sigma_1 = T$ and $\sigma_2 = \sigma_3 = 0$ at yielding. Substituting into equation 14.1,

$$2T^2 + K_1T^2 = K_2. \quad 14.3$$

In a uniaxial compression test in the 3 – direction, $\sigma_3 = -C$ and $\sigma_1 = \sigma_2 = 0$ at yielding. Substituting into equation 14.1,

$$2C^2 - K_1C^2 = K_2. \qquad 14.4$$

Combining, $2C^2 - K_1 C = 2T^2 + K_1T^2$, $K_1(T^2 + C^2) = 2(C^2 - T^2)$. $K_1 = 2(C^2 - T^2)/(C^2 + T^2)$.

$$\begin{aligned} K_2 &= 2T^2 + K_1T^2 = 2T^2 + T^2[2(C^2 - T^2)/(C^2 + T^2)] \\ &= (2T^4 + 2T^2C^2 + 2T^2C^2 - 2T^4)/(T^2 + C^2) \\ &= 4T^2C^2/(T^2 + C^2). \end{aligned} \qquad 14.5$$

Note, that if $T = C$, $K_1 = 0$ and $K_2 = 2T^2 = 2Y^2$, so equation 14.5 reduces to the von Mises criterion.

The predictions of equations 14.1 and 14.2 are very similar.

The Coloumb-Mohr criterion [3] is a pressure-modified Tresca criterion,

$$\sigma_1 - \sigma_3 + m(\sigma_1 + \sigma_2 + \sigma_3) = 2\tau_u, \qquad 14.6$$

where $2\tau_u$ is the yield strength in pure shear. The yield locus corresponding to this is plotted in Figure 14.8.

The constants, m and $2\tau_u$, in equation 14.6 can be expressed in terms of C and T. For a uniaxial tensile test in the 1-direction, $\sigma_1 = T$ and $\sigma_2 = \sigma_3 = 0$ at yielding so $T + mT = 2\tau_u$. At yielding in a compression test, $\sigma_3 = -C$ and $\sigma_1 = \sigma_2 = 0$, so $+C - mC = 2\tau_u$. Combining, $T + m\mathrm{Y} = +C - mC$, so $m = (C - T)/(C + T)$. $22\tau_u = T(1 + m) = T[1 + (C - T)/(C + T)] = 2CT/(C + T) = 2T/(C/T + 1)$.

The fundamental flow rule, $d\varepsilon_{ij} = d\gamma(\partial f/\partial\sigma_{ij})$ (equation 3.18), predicts that that if the yield criterion is sensitive to the level of hydrostatic pressure, $-(\sigma_1 + \sigma_2 + \sigma_3)/3$, yielding must be accompanied by a volume change. If $C > T$, the yielding should cause a volume increase. Figure 14.9 is a plot of volume strain, $\Delta v/v = (\varepsilon_1 + \varepsilon_2 + \varepsilon_3)$, against elongation for two polymers.

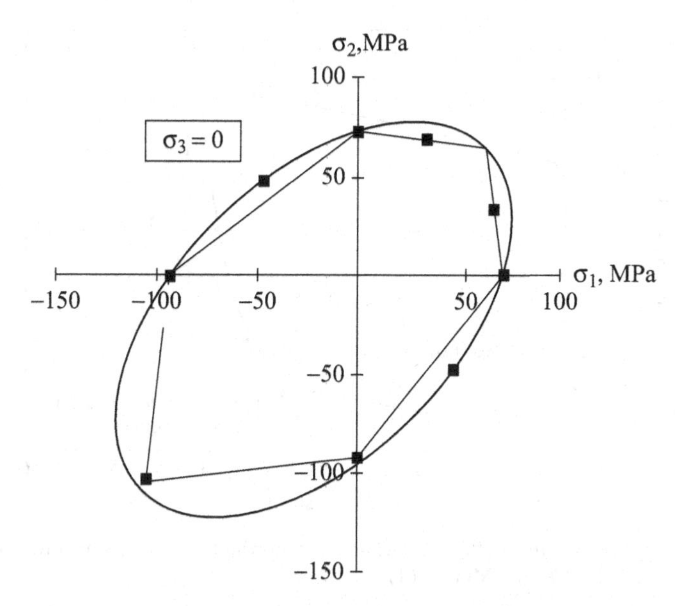

Figure 14.8. Pressure-dependent modifications of the von Mises (equation 14.4) and Tresca (equation 14.6) yield criteria for polystyrene. From W. F. Hosford, *Mechanical Behavior of Materials*, 2nd ed., Cambridge University Press (2010). Data from Whitney and Andrews, J. Polymer Sci., c-16 (1967).

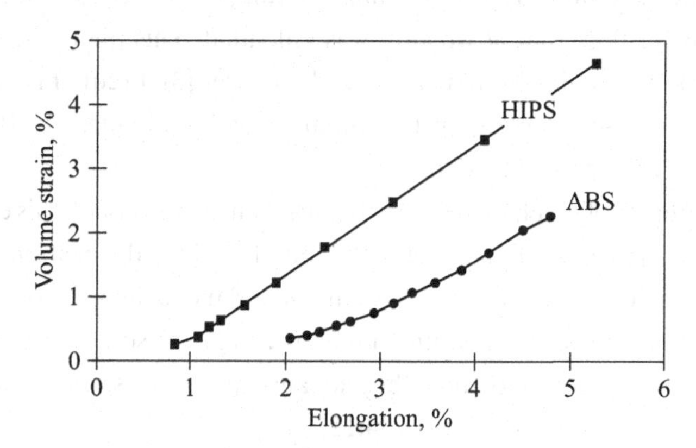

Figure 14.9. Volume change accompanying elongation. From W. F. Hosford, *Mechanical Behavior of Materials*, 2nd ed., Cambridge University Press (2010). Data from Bucknall, Partridge and Ward, *J. Mat. Sci.*, v. 19 (1984).

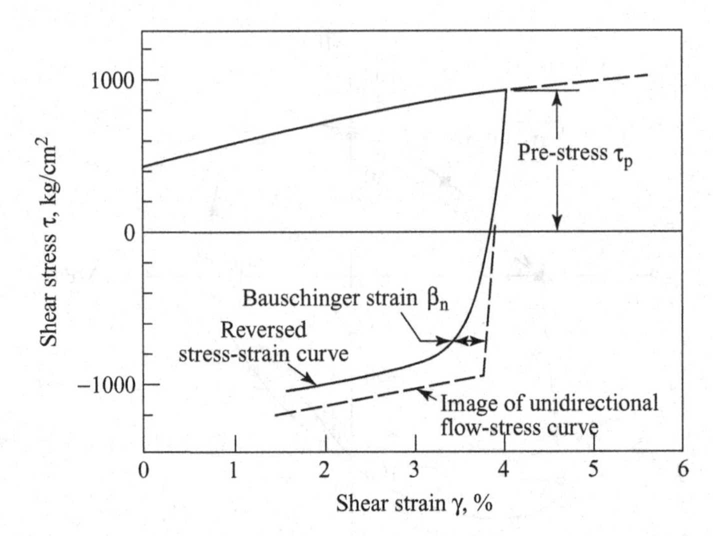

Figure 14.10. Bauschinger effect in steel under torsion and reverse torsion. From reference G. Deak, *ScD Thesis*, MIT (1961).

BAUSCHINGER EFFECT

If a material is plastically deformed under one type of loading and then subjected to another type of loading, yielding usually commences at a stress lower than that at which it was unloaded. This phenomenon is called the Bauschinger effect after its discoverer [3]. Frequently, there is some reverse straining on the initial unloading. Figures 14.10 and 14.11 are examples of this effect.

A number of mechanisms have been advanced to explain this effect. Orowan [4] proposed that during the initial loading, dislocations pile-up against obstacles such as a grain boundary, a hard particle, or dislocation forrest; that on unloading, they retreat somewhat by the back stress; and, on reloading, they advance at a lower stress until they meet new obstacles (see Figure 14.12).

Different yield strengths of different regions, A and B, will cause a Bauschinger effect. Figure 14.13 shows the stress strain curves of two neighboring regions of a material. They must undergo the same

Stress
Strain

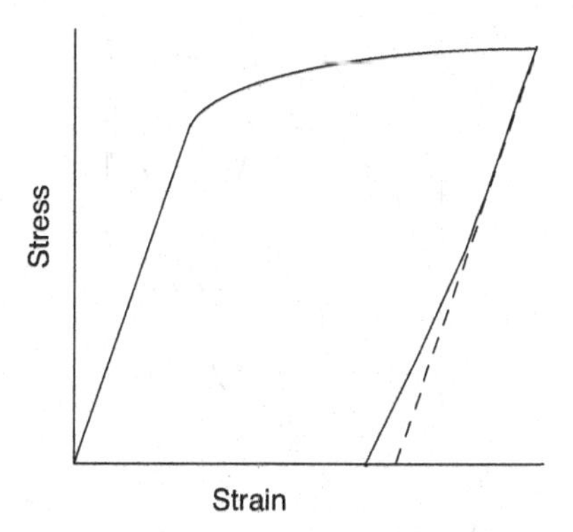

Figure 14.11. Reverse yielding during unloading lowering. From W. F. Hosford and R. M. Caddell, *Metal Forming; Mechanics and Metallurgy*, 4th ed. Cambridge University Press (2011).

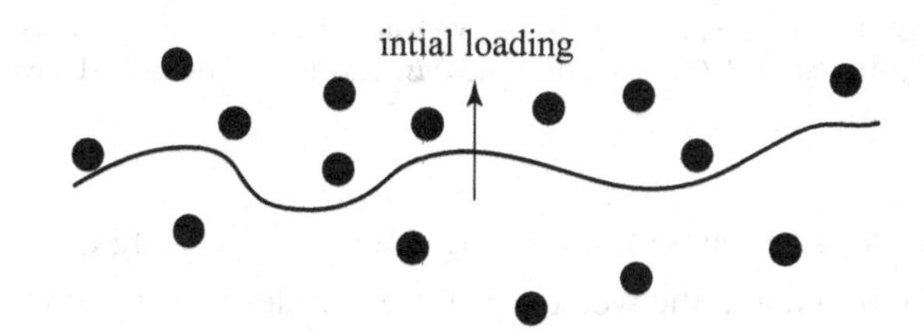

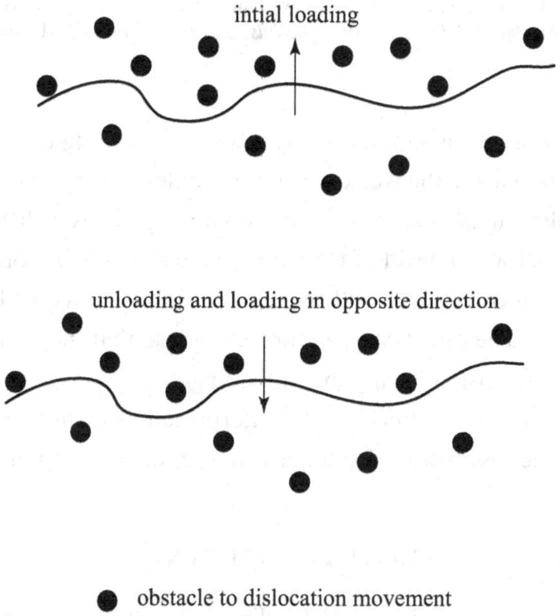

intial loading
unloading and loading in opposite direction
obstacle to dislocation movement

Figure 14.12. Dislocation model for the Bauschinger effect.

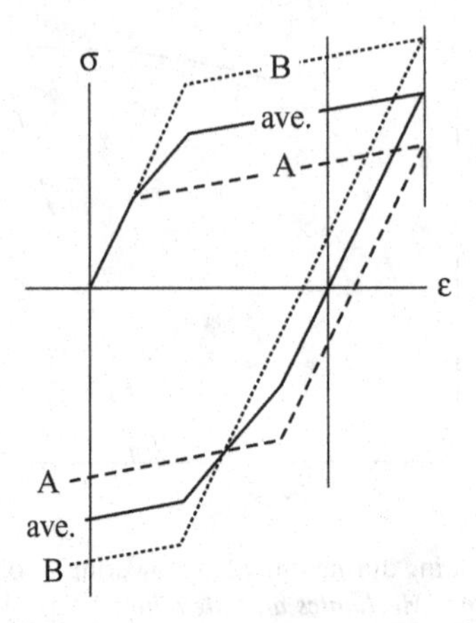

Figure 14.13. Different stress strain curves of two adjacent grains will result in a Bauschinger effect. Both grains are constrained to undergo the same strains. After unloading, grain A will be left under residual compression when the average stress is zero. On loading in compression it will yield while the compressive stress in grain B is very small. The average stress-strain curve will show a Bauschinger effect. From W. F. Hosford, *Mechanical Behavior of Materials*, 2nd ed., Cambridge University Press (2010).

strains under tension and unloading. During unloading the same elastic contraction leave the weaker grain, A, under residual compression. During subsequent loading in compression, grain A will yield at a stress well below the level of the average tensile stress before unloading. The composite curve will therefore exhibit a low yield stress in compression. The effect can be large enough so that the weaker grain yields in compression during the unloading.

The Bauschinger effect can be interpreted as a shift in the yield locus. See the discussion of kinematic hardening in Chapter 12.

NOTE OF INTEREST

William C. Leslie had an outstanding thirty-five year career, first as an industrial research engineer and, subsequently, since 1973, as teacher and researcher at The University of Michigan. Professor Leslie received his BS, MSc, and PhD. degrees, all in metallurgical engineering, from Ohio State University in 1947, 1948, and 1949, respectively. His career started as a metallurgist at the United States Steel research laboratory in Kearny, New Jersey, during the period of 1949 to 1953. During 1952 to 1953, he was also an adjunct professor at Brooklyn Polytechnic Institute. In 1953, he became associate director of research and development at Thompson Products, Inc., Cleveland, Ohio. He returned to the U.S. Steel Corporation research laboratory, at the Bain Laboratory for Fundamental Research, in 1953, and remained there until 1973.

He joined the faculty of The University of Michigan in 1973. He was an outstanding contributor throughout his career in the area of the physical metallurgy of steels. He wrote nearly 100 research papers and has authored the text, *The Physical Metallurgy of Steels*, which was published in 1981. He retired from active faculty status as of December 31, 1984.

Johann Bauschinger was born June 11, 1834 in Nurnberg Germany and died on November 25, 1893 in Munich. He was a mathemetician, a builder, and professor of Mechanics at Munich Polytechnic from 1868 until his death.

Egon Orowan was born in Budapest on August 2, 1902. In 1928, he enrolled at the Technical University of Berlin. His doctorate, in 1932, was on fracture of mica. In 1934, Orowan along with Taylor and Polanyi proposed the of the theory of dislocations. In 1937, he worked on fatigue at the University of Birmingham. In 1939, he moved to University of Cambridge where he worked on problems of munitions production, particularly that of plastic flow during rolling. In 1944, he became in understanding the causes of the tragic loss of many Liberty ships during the war, identifying the poor notch sensitivity of welds and the aggravating effects of the extreme low temperatures. In 1950, he moved to MIT where, in addition to continuing his metallurgical work, he developed his interests in geological fracture.

REFERENCES

1. W. C. Leslie, *The Physical Metallurgy of Steels*, Hemisphere Publishing Company (1981).
2. C. A. Coulomb, "Essai sur une application des regles des maximis et minimis a quel problemes de statique relatifs, a la architecture." *Mem. Acad. Roy. Div. Sav.* v. 7 (1776).
3. J. Bauschinger, *Mitteilung aus dem Mechanisch, Technischen Laboratorium der K.TechnischebHochshule in Munchen* v. 13 (1886).
4. E. Orowan. In *Internal Stresses and Fatigue in Metals*, General Motors Symposium (1959).

15

LOWER-BOUND MODELS

LOWER-BOUND AVERAGING

Lower bounds are quite different than upper bounds [1]. They are based on self-consistent internal stress fields, without regard to assuring self-consistent internal deformation patterns. The difference is illustrated in Figure 15.1 for uniaxial tension. Averaging the strains at the same stress level (a) in all grains results in a lower bound whereas averaging the stresses at the same strain (b) results in an upper bound.

Lower-bound averages are reasonable only if strain hardening is assumed. Otherwise, the stress cannot rise above the yield strength of the weakest grain so this becomes the average. This is illustrated for uniaxial tension in Figure 15.2. Therefore, strain hardening must be considered in calculating lower bounds.

One simple way of making lower-bound calculations is to assume power-law hardening.

$$\tau_i = k\gamma_i^n, \tag{15.1}$$

where τ_i is the shear stress necessary to cause slip in grain, *i*, and γ_i is the shear strain in grain, i. Another basic assumption is that yielding is defined to occur when the plastic work in the polycrystal reaches a critical level, w^*. This is equivalent to defining yielding in terms of an effective strain, $\bar{\varepsilon}$, because $\bar{\sigma}$ and $\bar{\varepsilon}$ are constant everywhere on the yield surface and $w = \int \bar{\sigma}\, d\bar{\varepsilon}$. It is also assumed that only the most

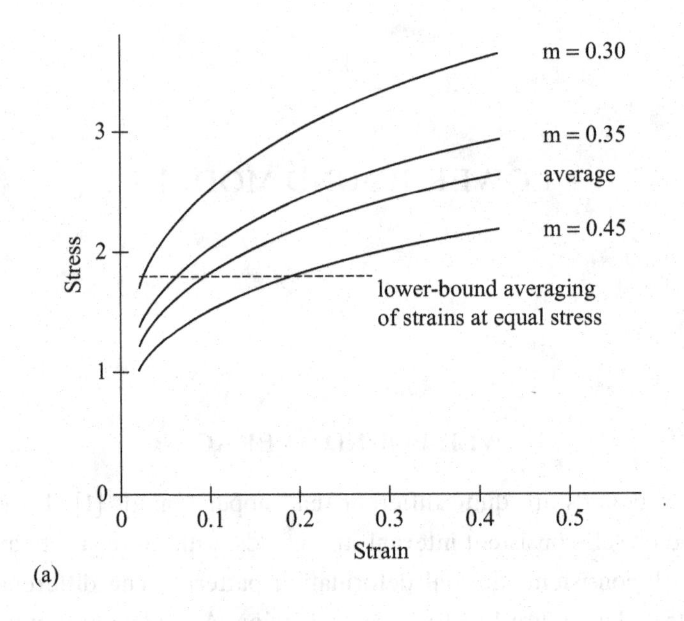

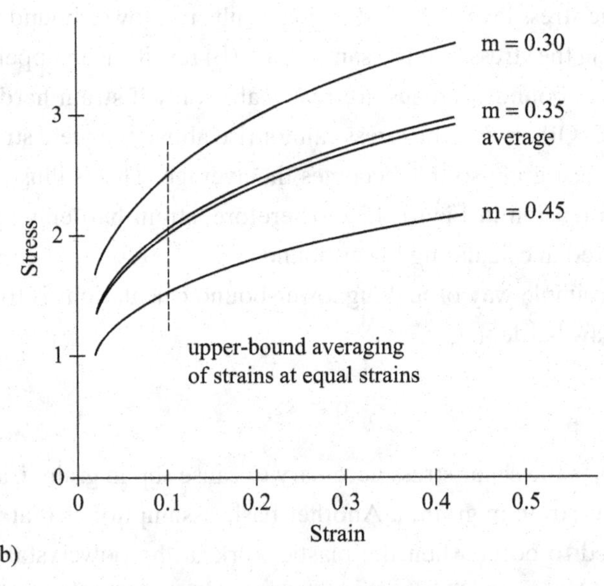

Figure 15.1. Stress strain curves for polycrystals consisting of three grains, each with a different stress strain curve. A lower bound (a) results from averaging the strains at the same stress level. An upper bound results from averaging the stresses at the same strain level. From W. F. Hosford and A. Galdos, *Textures and Microstructure* v. 12 (1990).

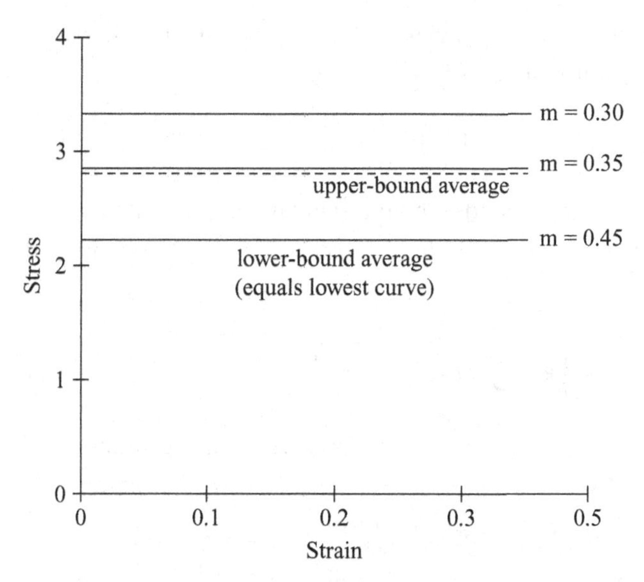

Figure 15.2. Upper- and lower-bound averaging without strain hardening. Note the lower bound average must be the level of the lowest curve. From W. F. Hosford and A. Galdos, *Textures and Microstructure* v. 12 (1990).

highly stressed slip system in each grain operates and a state of plane stress, σ_x and $\sigma_x = \alpha\sigma_x$, with $\sigma_z = \tau_{yz} = \tau_{zx} = \tau_{xy} = 0$. The plastic work per volume, w_i, in an individual grain is given by

$$w_i = \tau_i d\gamma_i = k\gamma_i^{n+1}/(n+1) = k^{-1/n+1}\tau_i^{(n+1)/n}/(n+1), \qquad 15.2$$

where τ_i and γ_i are the shear stress and shear strain on the most heavily stressed slip system in grain i. For a polycrystal containing g grains of equal size, the work per volume, w is

$$w = \left(\frac{q}{g}\right)\sum \tau_i^{(n+1)/n}, \qquad 15.3$$

where $q = 1/[(n+1)k^{1/n}]$. At yielding, $w = w^*$. For a given stress ratio, α, the shear stress, τ_i is

$$\tau_i = m_{xi}\sigma_x + m_{yi}\sigma_y = \sigma_x(m_{yi} + \alpha m_{yi}). \qquad 15.4$$

Here the resolving factor is $m_{xi} = \cos\lambda_i\cos\phi_i$, where λ_i and ϕ_i are the angles between the x-direction and the slip direction and the slip plane normal. The resolving factor, m_{xi} is defined similarly. Substituting equation 15.4 into equation 15.3,

$$w^* = \left(\frac{q}{g}\right)\sum[\sigma_x(m_{xi} + \alpha m_{xi})^{n/(n+1)}. \qquad 15.5$$

Because σ_x is assumed to be the same in all grains, it can be removed from the summation. Taking $w^*/q = w^*k^{1/n}(n+1)$ as unity and solving for σ_x,

$$\sigma_x = \left[g/\sum_i (m_{xi} + \alpha m_{yi})^{n+1)/n'}\right]^{n/(n+1)} \quad \text{and } \sigma_y = \alpha\sigma_x \qquad 15.6$$

The ratio of the resulting strains, $\varepsilon_y/\varepsilon_x$, can be found in a similar manner. In each grain, $\varepsilon_{ix} = m_{ix}\gamma_i$ and $\varepsilon_{iy} = m_{iy}\gamma_i$, where $\gamma_i = (\tau_i/k)^{1/n}$ and $\tau_i = \sigma_x(m_{yi} + \alpha m_{yi})$. Substituting,

$$\varepsilon_{ix} = m_{ix}\left(\sigma_x\frac{m_{ix} + \alpha m_{iy}}{k}\right)^{1/n}$$

$$\varepsilon_{iy} = m_{iy}\left(\sigma_y\frac{m_{ix} + \alpha m_{iy}}{k}\right)^{1/n} \qquad 15.7$$

Therefore, the ratio of the external strains, $\varepsilon_x = (1/g)\Sigma\varepsilon_{ix}$ and $\varepsilon_y = (1/g)\Sigma\varepsilon_{iy}$, is

$$\varepsilon_x/\varepsilon_y = \Sigma\varepsilon_{ix}/\Sigma\varepsilon_{iy} \qquad 15.8$$

Lower-bound calculations for {111}<110> slip were made for 100 randomly selected textures, each consisting five randomly chosen sheet normals with 15 different rotations about the normal [1]. For each texture, the ratios of stresses β, λ and X defined in Chapter 9 as well as the R-value. The results of these calculations are shown in Figures 15.3 through 15.7.

The Correlation coefficient, R^2 for various levels of the exponent a in equation 7.30 are plotted in Figures 15.6 through 15.10. In all cases, an exponent $a = 8$ gives a good fit.

Similar lower bound calculations have been made for <111> pencil-glide. The results of these calculations are shown in

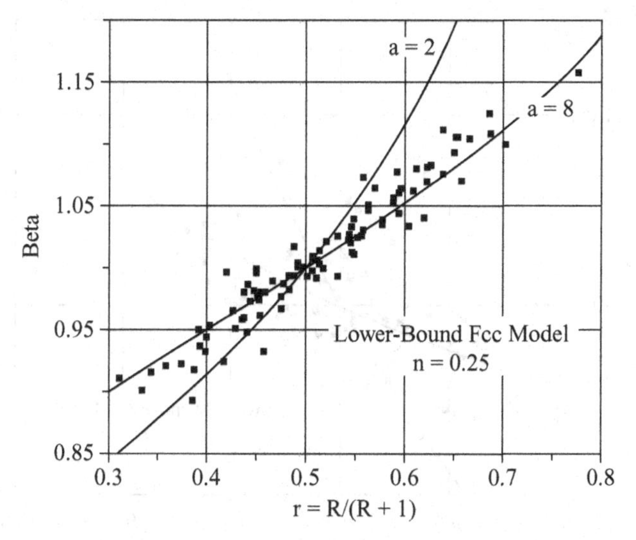

Figure 15.3. Calculations of β and R for lower-bound fcc metals with $n = 0.25$. From W. F. Hosford and A. Galdos, *Textures and Microstructure* v. 12 (1990).

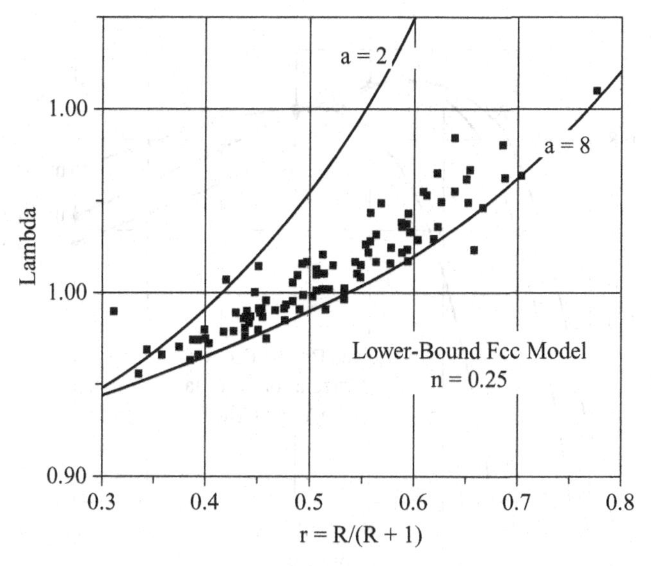

Figure 15.4. Calculations of λ and R for lower-bound fcc metals with $n = 0.01$. From W. F. Hosford and A. Galdos, *Textures and Microstructure* v. 12 (1990).

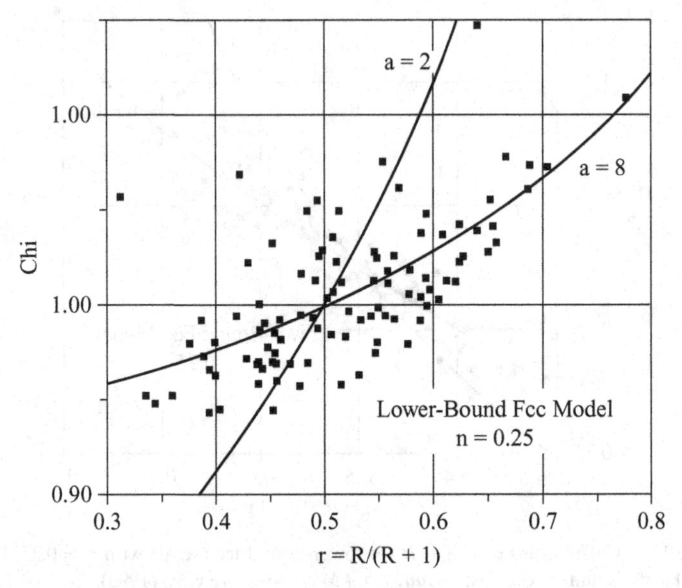

Figure 15.5. Calculations of X and R for lower-bound fcc metals with $n = 0.25$. From W. F. Hosford and A. Galdos, *Textures and Microstructure* v. 12 (1990).

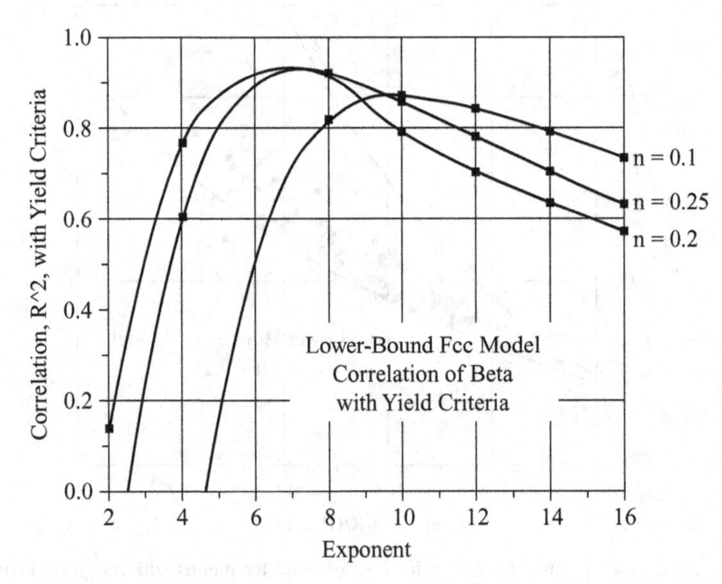

Figure 15.6. Correlation of β exponent, a. From W. F. Hosford and A. Galdos, *Textures and Microstructure* v. 12 (1990).

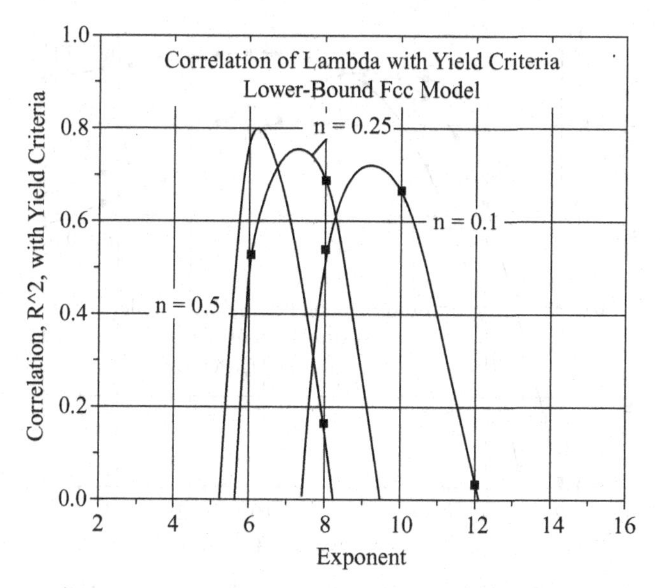

Figure 15.7. Correlation of λ exponent, *a*. From W. F. Hosford and A. Galdos, *Textures and Microstructure* v. 12 (1990).

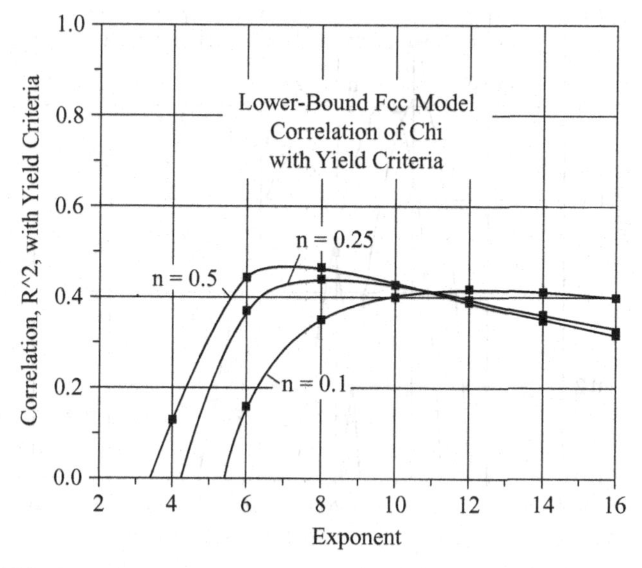

Figure 15.8. Correlation of χ exponent, *a*. From W. F. Hosford and A. Galdos, *Textures and Microstructure* v. 12 (1990).

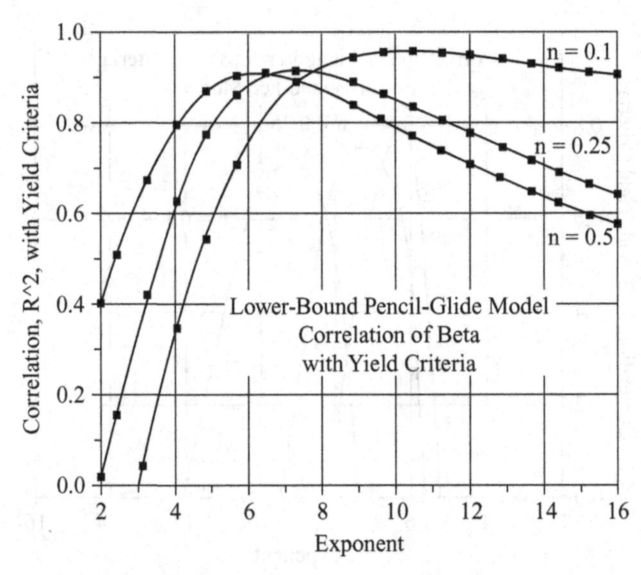

Figure 15.9. Correlation of β exponent, a. From W. F. Hosford and A. Galdos, *Textures and Microstructure* v. 12 (1990).

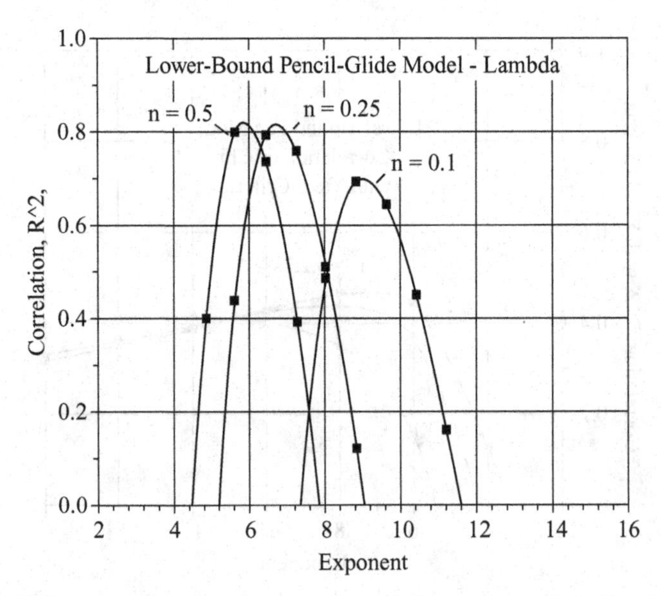

Figure 15.10. Correlation of λ exponent, a. From W. F. Hosford and A. Galdos, *Textures and Microstructure* v. 12 (1990).

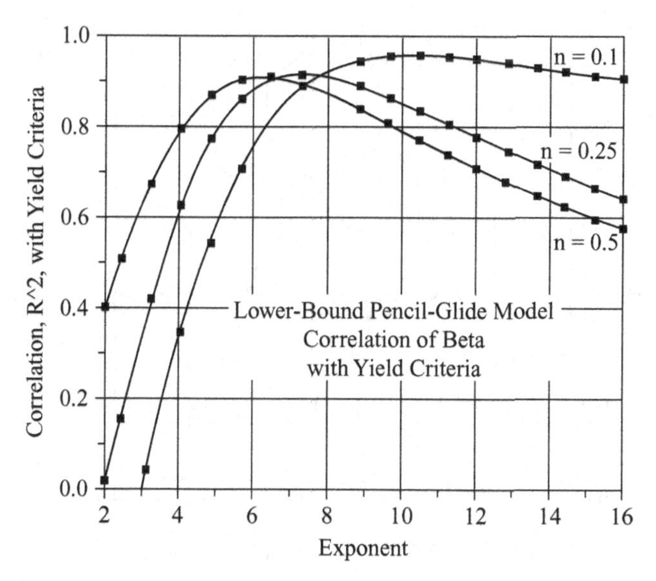

Figure 15.11. Fit of exponent a for β from lower bound pencil glide model. From W. F. Hosford and A. Galdos, *Textures and Microstructure* v. 12 (1990).

Figures 15.11 and 15.12. Here, except for $n = 0.1$, the best fit occurs near $a = 6$.

Figure 15.13 shows the exponents, a, in equation 7.30 that result in the best fit of the lower bound calculations for values of λ. The best fit levels decreases as the strain hardening exponent increases.

OTHER MODELS FOR LOWER BOUNDS

The term "lower bound" has been applied incorrectly to models of polycrystalline strength that do not fulfill the lower-bound requirement of a statically admissible stress field [2–7]. Sachs [2] related the tensile yield strength of a randomly oriented polycrystal to the shear stress for slip by averaging the reciprocal Schmid factors, $1/m$, and found that $\theta/\tau = (1/m)_{\text{av}} = 2.24$. Although this implies a different stress in each grain, it can be regarded as a lower bound for a polycrystal in tension only if all grain boundaries are parallel to the to the tensile axis. Otherwise, there would be unbalanced stresses across

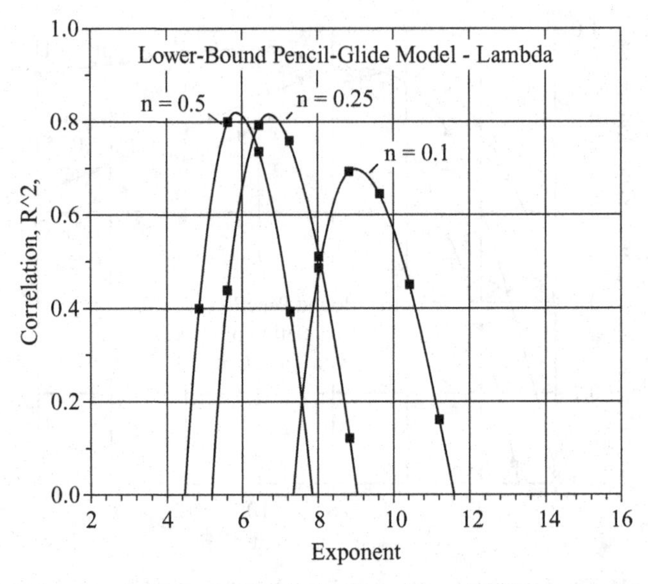

Figure 15.12. Fit of exponent *a* for λ from lower bound pencil glide model. From W. F. Hosford and A. Galdos, *Textures and Microstructure* v. 12 (1990).

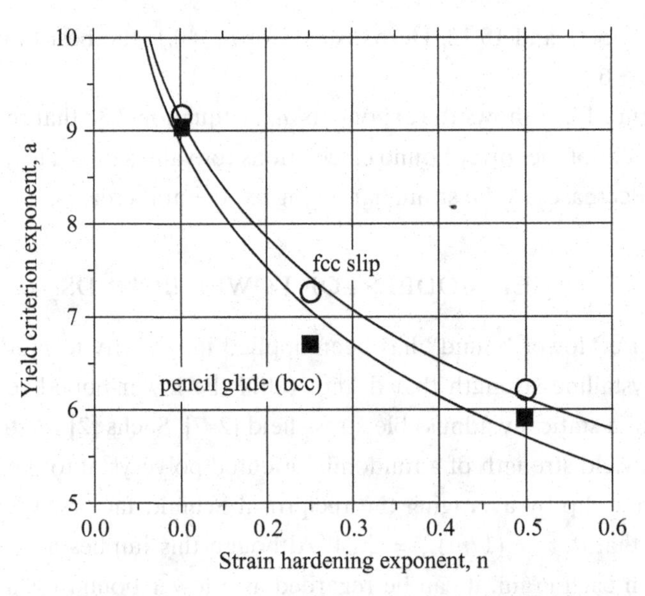

Figure 15.13. Yield criterion exponents, a, that result in the best fits for lower bound calculations for values of λ. From W. F. Hosford, *The Mechanics of Crystals and Textured Polycrystals*, 2nd ed., Cambridge University Press (1993).

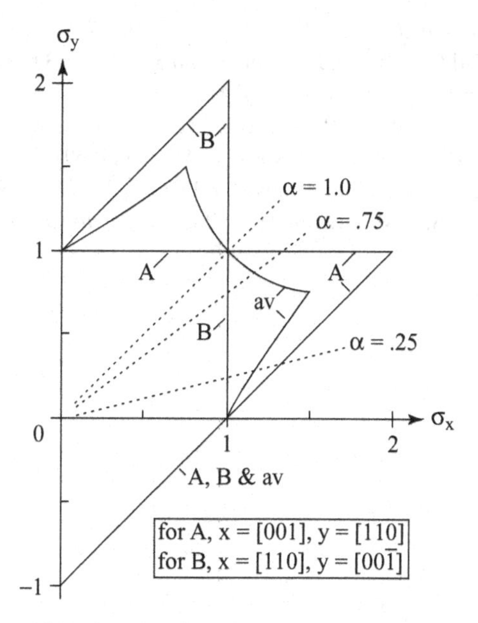

Figure 15.14. Illustration of a violation of normality when stresses in grains loaded under the same stress ratio, but different magnitudes are averaged. Note the outward concavity. From W. F. Hosford and A. Galdos, *Textures and Microstructure* v. 12 (1990).

grain boundaries. There have been several attempts to extend the Sachs' model by averaging the stresses in all grains that are loaded under the same stress ratio, α. However, this is not a lower bound and violates the normality principle as shown in Figure 15.14.

NOTE OF INTEREST

George Sachs was born in Moscow on April 5, 1896. He taught at Frankfurt University from 1930 to 1935 and at Case University. He was of Jewish birth, and left Germany with his family in 1937 to escape Nazi persecution. He died on October 30, 1960.

REFERENCES

1. W. F. Hosford and A. Galdos, *Textures and Microstructure* v. 12 (1990).
2. W. F. Hosford, *The Mechanics of Crystals and Textured Polycrystals*, Oxford University Press (1993).

3. G. Sachs, *Zeitschrift Verein Deut. Ing.* v. 72 (1928).
4. J. Althoff and P. Wincerz, *Zeits. für Metallkunde*, v. 63 (1972).
5. W. F. Hosford, *Met. Trans.* v. 5 (1974).
6. H. R. Piehler and W. A. Backofen. In *Textures in Research and Practice*, Wasserman and Grewen eds., Springer Verlag (1969).
7. C. S. Da Viana, J. S. Kallend, and G. J. Davies, *I. J. Mech. Sci* v. 21 (1979).
8. S. L. Semiatin, S. L. Morris, and H. R. Piehler, *Texture of Crystalline Solids*, v. 3 (1979).

16

PLASTICITY TESTS

Most data on plasticity are derived from tensile tests. Stress-strain data may also be derived from other tests.

TENSION TESTS

Necking, as described in Chapter 13, starts when $d\sigma/d\varepsilon = \sigma$. Once necking starts, the stress in the necked region is no longer one of uniaxial tension. As material in the center of the neck is being stretched in the axial direction, it must contract laterally. This contraction is resisted by the adjacent regions, immediately above and below, that have larger cross sections and are therefore not deforming. The net effect is that the center of a neck is under triaxial tension.

Figure 16.1 shows the stress distribution calculated by Bridgman [1].

Only that part of the axial stress which exceeds the lateral stress is effective in causing yielding. Bridgman showed that the effective part of the stress, $\bar{\sigma}$ is

$$\bar{\sigma}/\sigma = 1/\{(1 + 2R/a)\ln[1 + a/(2R)]\}, \qquad 16.1$$

where σ is the measured stress, a is the radius of specimen at the base of the neck, and R is the radius of curvature of the neck profile.

Figure 16.2 is a plot of the Bridgman correction factor, $\bar{\sigma}/\sigma$, as a function of a/R according to equation 16.1. A simple way of measuring

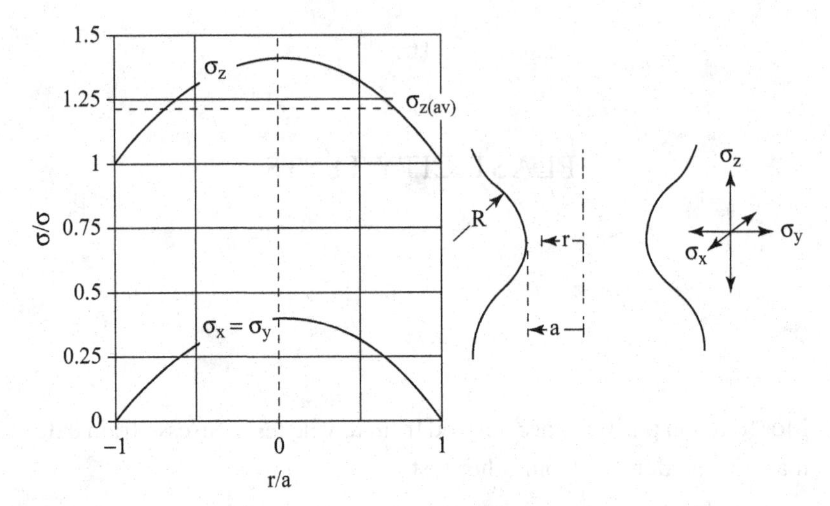

Figure 16.1. The stress distribution across a neck (left) and the corresponding geometry of the neck (right). Both axial and lateral tensile stresses are a maximum at the center of the neck. From W. F. Hosford, *Mechanical Behavior of Materials*, 2nd ed., Cambridge University Press (2011).

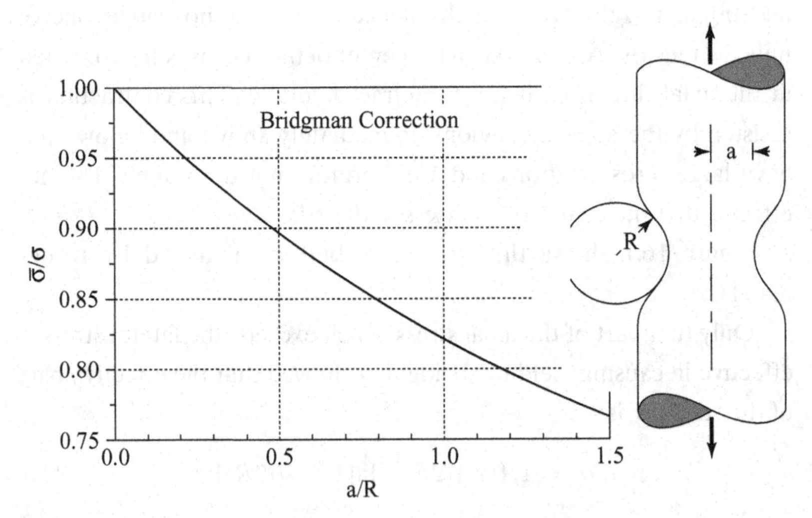

Figure 16.2. The Bridgman correction factor as a function of the neck shape. The plot gives the ratio of the effective stress to the axial true stress for measured values of a/R. From W. F. Hosford, *Mechanical Behavior of Materials*, 2nd ed., Cambridge University Press (2011).

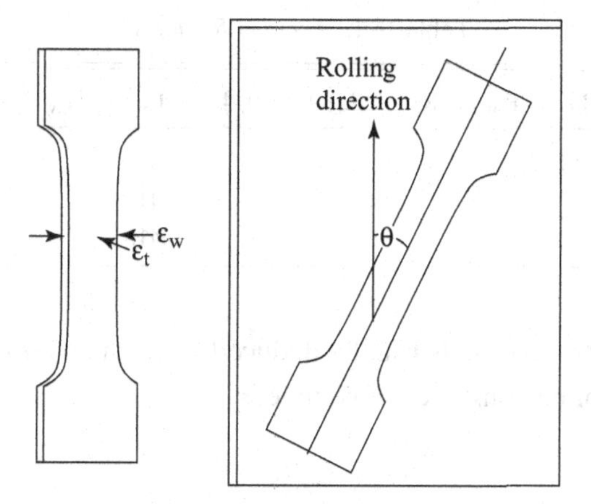

Figure 16.3. Tensile specimen cut from a sheet used to determine the R-value. From W. F. Hosford, *Mechanical Behavior of Materials*, 2nd ed., Cambridge University Press (2011).

the radius of curvature, R, can be measured by sliding a calibrated cone along the neck until it becomes tangent at the base of the neck.

SHEET ANISOTROPY

The angular variation of yield strength in many sheet materials is not large. However, such a lack of variation does not indicate that the material is isotropic. The parameter that is commonly used to characterize the anisotropy is the *strain ratio* or *R-value*[1] (Figure 16.3). This is defined as the ratio, R, of the contractile strain in the width direction to that in the thickness direction during a tension test,

$$R = \varepsilon_w/\varepsilon_t. \qquad 16.2$$

during a tension test. In an isotropic material, the width and thickness strains, ε_w and ε_t, are equal so the value of R for an isotropic material is 1. However, R is usually either greater or less than one in real sheet materials. Direct measurement of the thickness strain in thin sheets is

[1] Some authors use the symbol, r, instead of R.

Table 16.1. *Average R values*

R_0	R_{45}	R_{90}	$R_{av} = (R_0 + 2R_{45} + R_{90})/4$	J_{av}	$R_{av} = -J_{av}/(1 - J_{av})$
2	0.5	2	1.25	0.5	1.0
2	0.5	1	1.0	0.417	0.715
1	0.5	1	0.75	0.418	0.718

inaccurate. Instead, ε_t is usually deduced from the width and length strains assuming constancy of volume, $\varepsilon_t = -\varepsilon_w - \varepsilon_l$.

$$R = -\varepsilon_w/(\varepsilon_w + \varepsilon_l) \qquad 16.3$$

The *R*-value usually varies with the direction of testing. Commonly an average *R* value is taken as

$$R = (R_0 + R_{90} + 2R_{45})/4 \qquad 16.4$$

However, that way of averaging *R*-values put a greater emphasis on high value than on low values. For example, an *R*-value of ½ is just as extreme as an *R*-value of 2, but the average of ½ and 2 is 1.25. It would be better to average a function $J = R/(R+1)$. The $J_{av} = [R_0/(R_0+1) + 2R_{45}/(R_{45}+1) + R_{90}/(R_{90}+1)]/4$ and $R_{av} = J_{av}/(1 - J_{av})$. Table 16.1 illustrates the difference in the two way of averaging.

The angular variation is usually characterized by

$$\Delta R = (R_0 + R_{90} - 2R_{45})/2, \qquad 16.5$$

although it would probably be better to characterize it with $\Delta J = (J_0 + J_{90} - 2\,J_{45})/2$.

To avoid constraint from the shoulders, strains should be measured well away from the ends of the gauge section. Some workers suggest that the strains be measured when the total elongation is 15 percent, if this is less than the necking strain. The change of *R* during a tensile test is usually quite small and the lateral strains at 15 percent elongation are great enough to be measured with accuracy.

DEFINITION OF YIELDING

Yielding in other types of tests must be carefully defined if it is to be used with plasticity theory. If yielding in a tension test is defined by a 0.2 percent offset, for the purpose of assessing the anisotropy, yielding under any other form of loading should be defined by the plastic strain that involves the same amount of plastic work per volume as 0.2 percent offset in tension.

COMPRESSION TEST

Necking limits the uniform elongation in tension. Much higher strains are achievable in compression tests. However, there are two problems that limit the usefulness of compression tests: friction and buckling. Friction on the ends of the specimen tends to suppress the lateral spreading of material near the ends (Figure 16.4). A cone-shaped region of *dead metal* (undeforming material) can form at each end with the result that the specimen becomes barrel shaped. Friction can be reduced by lubrication and the effect of friction can be lessened by increasing the height-to-diameter ratio, h/d, of the specimen.

If the coefficient of friction, μ, between the specimen and platens is constant, the average pressure to cause deformation is

$$P_{av} = Y(1 + (\mu d/h)/3 + (\mu d/h)^2/12 + \cdots), \qquad 16.6$$

where Y is the true flow stress of the material. If, on the other hand, there is a constant shear stress at the interface, such as would be obtained by inserting a thin film of a soft material (for example, lead, polyethylene, or Teflon), the average pressure is

$$P_{av} = Y + (1/3)k(d/h), \qquad 16.7$$

where k is the shear strength of the soft material. However, these equations usually do not accurately describe the effect of friction because

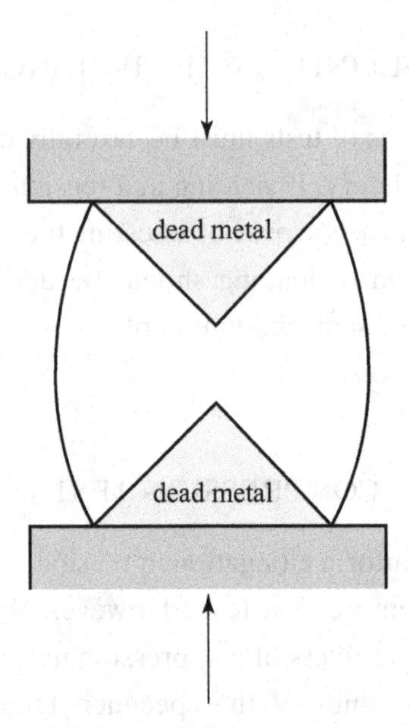

Figure 16.4. Unless the ends of a compression specimen are well lubricated, there will be a conical region of undeforming material (dead metal at each end of the specimen. As a consequence the mid section will bulge out or *barrel*. From W. F. Hosford, *Mechanical Behavior of Materials*, 2nd ed., Cambridge University Press (2011).

neither the coefficient of friction nor the interface shear stress is constant. Friction is usually highest at the edges where liquid lubricants are lost and thin films may be cut during the test by sharp edges of the specimens. Severe barreling caused by friction may cause the sidewalls to fold up and become part of the ends as shown in Figure 16.5. Periodic unloading to replace the film or relubricate helps reduce these effects.

Although increasing h/d reduces the effect of friction, the specimen will buckle if it is too long and slender. Buckling is likely if the height-to-diameter ratio is greater than about 3. If the test is so well lubricated that the ends of the specimen can slide relative to the platens, buckling can occur for $h/d \geq 1.5$ (Figure 16.6).

Problems with compression testing: a) Friction at the ends prevents spreading which results in barreling; b) buckling of poorly lubricated specimens can occur if the height-to- diameter ratio, h/d, exceeds

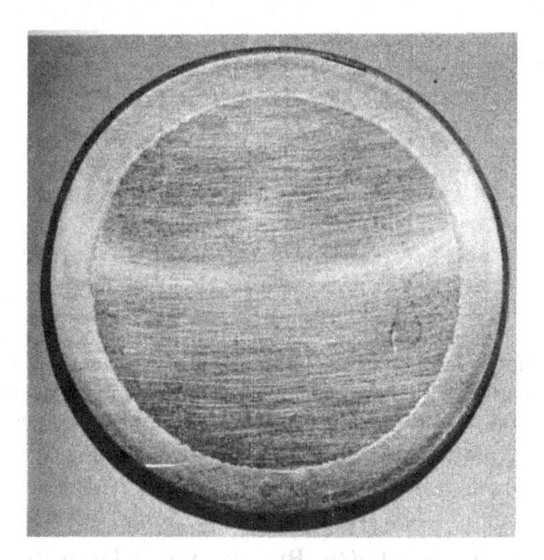

Figure 16.5. Photograph of the end of a compression specimen. The darker central region was the original end. The lighter region outside was originally part of the cylindrical wall that folded up with the severe barreling. From G. W. Pearsall and W. A. Backofen, *Journal of Engineering for Industry, Trans ASME* v. 85B (1963).

about 3. Without any friction at the ends (c), buckling can occur if h/d is greater than about 1.5.

One way to overcome the effects of friction [4] is to test specimens with different diameter/height ratios. The strains at several levels of

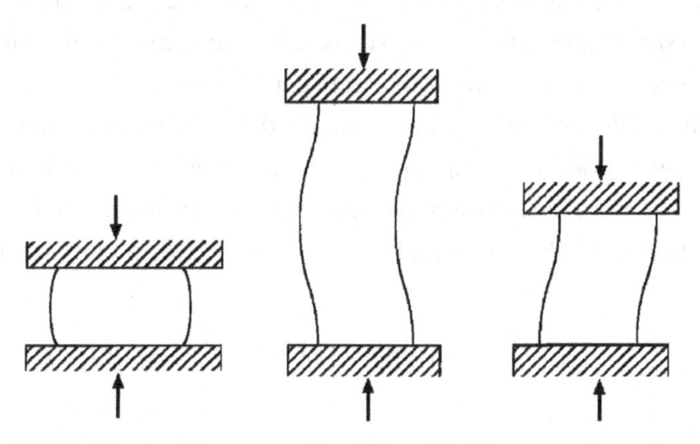

Figure 16.6. Problems with compression testing: (a) Friction at the ends prevents spreading which results in barreling; (b) buckling of poorly lubricated specimens can occur if the height-to- diameter ratio, h/d, exceeds about 3. Without any friction at the ends (c) buckling can occur if h/d is greater than about 1.5. From W. F. Hosford, *Mechanical Behavior of Materials*, 2nd ed., Cambridge University Press (2011).

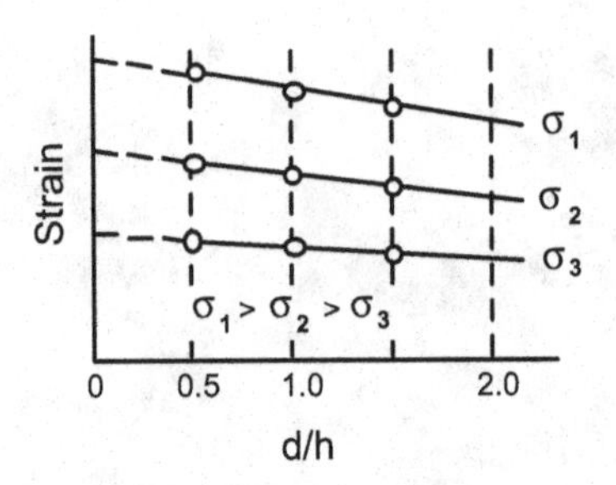

Figure 16.7. Extrapolation scheme for eliminating frictional effects in compression testing. Strains at different levels of stress ($\sigma_1, \sigma_2, \sigma_3$) are plotted for specimens of differing heights. The strain for "frictionless" conditions is obtained by extrapolating d/h to 0. From W. F. Hosford, *Mechanical Behavior of Materials*, 2nd ed., Cambridge University Press (2011).

stress are plotted against d/h. By the extrapolating the stresses to $d/h = 0$, the stress levels can be found for an infinitely long specimen in which the friction effects would be negligible (Figure 16.7).

During compression the cross-sectional area that carries the load increases. Therefore, in contrast to the tension test, the absolute value of engineering stress is greater than the true stress (Figure 16.8). The area increase, together with work hardening, can lead to very high forces during compression tests, unless the specimens are very small.

The shape of the engineering stress-strain curve in compression can be predicted from the true stress-strain curve in tension, assuming that absolute values of true stress in tension and compression are the same at the same absolute strain-values. In converting true stress and true strain to compressive stress and compressive strain, it must be remembered that both the stress and strain are negative in compression,

$$e_{comp} = \exp(\varepsilon) - 1, \tag{16.8}$$

and

$$s_{comp} = \sigma/(1 + e). \tag{16.9}$$

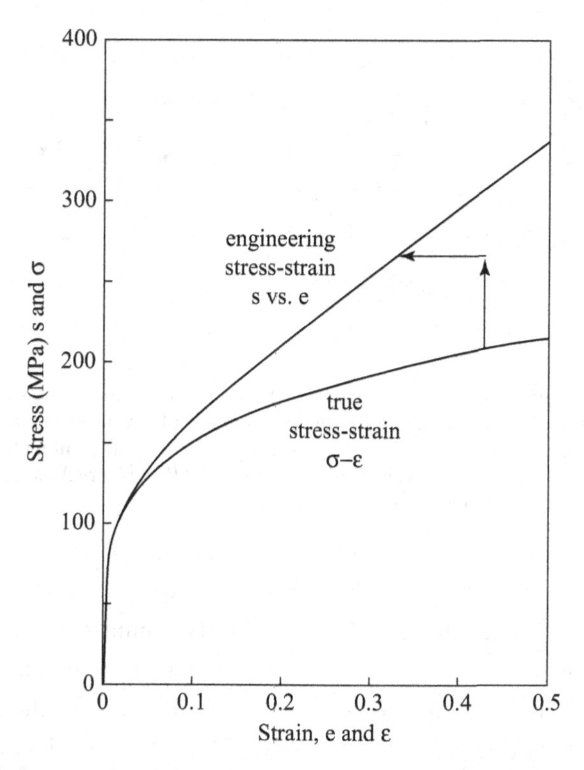

Figure 16.8. Stress strain relations in compression for a ductile material. Each point, s, e on the true stress- true strain curve corresponds to a point, *s*, *e*, on the engineering stress-strain curve. The arrows connect these points. From W. F. Hosford, *Mechanical Behavior of Materials*, 2nd ed., Cambridge University Press (2011).

PLANE-STRAIN COMPRESSION

There are two simple ways of making plane-strain compression tests. Small samples can be compressed in a channel that prevents spreading (Figure 16.9a). In this case, there is friction on the sidewalls of the channel as well as on the platens so the effect of friction is even greater than in uniaxial compression. An alternative way of producing plane-strain compression is to use a specimen that is very wide relative to the breadth of the indenter (Figure 16.9b). This eliminates the sidewall friction, but the deformation at and near the edges deviates from plane

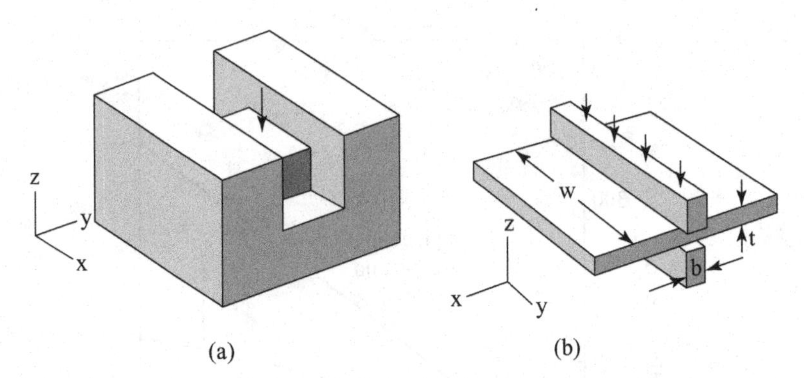

Figure 16.9. Plane strain compression tests; a) compression in a channel with the side walls preventing spreading, b) plane strain compression of a wide sheet with a narrow indenter. Lateral constraint forcing $\varepsilon_y = 0$, is provided by the adjacent material that is not under the indenter. From W. F. Hosford, *Mechanical Behavior of Materials*, 2nd ed., Cambridge University Press (2011).

strain. This departure from plane strain extends inward for a distance approximately equal to the indenter width. To minimize this effect, it is recommended that the ratio of the specimen width to indenter width, w/b, be about 8. It is also recommended that the ratio of the indenter width to sheet thickness, b/t, be about 2. Increasing b/t increases the effect of friction. Both of these tests simulate the plastic conditions that prevail during flat rolling of sheet and plate. They find their greatest usefulness in exploring plastic anisotropy.

PLANE-STRAIN TENSION

Plane-strain can be achieved in tension with specimens having gauge sections that are very much wider than they are long [2]. Figure 16.10 shows several possible specimens and specimen gripping arrangements. Such tests avoid all the frictional complications of plane-strain compression. However, the regions near the edges lack the constraint necessary to impose plane strain. At the very edge, the stress preventing contraction disappears so the stress state is uniaxial tension. Corrections must be made for departure from plane strain flow near the edges.

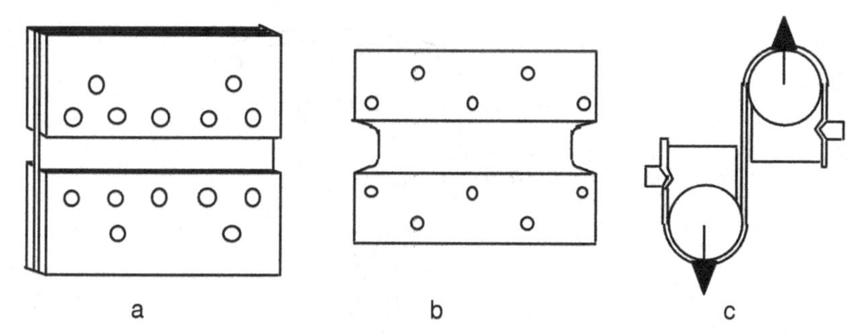

Figure 16.10. Several ways of making plane-strain tension tests on sheet specimens. All have a gauge section that is very short relative to its width: (a) Enlarged grips produced by welding to additional material. (b) Reduced gage section cut into edge. (c) Very short gage section achieved by friction on the cylindrical grips. From W. F. Hosford, *Mechanical Behavior of Materials*, 2nd ed., Cambridge University Press (2011).

BIAXIAL TENSION (HYDRAULIC BULGE TEST)

Much higher strains can be reached in bulge tests than in uniaxial tension tests [5, 6, 7]. This allows evaluation of the stress-strain relationships at high strains. A set-up for bulge testing is sketched in Figure 16.11. A sheet specimen is placed over a circular hole, clamped, and bulged outward by oil pressure acting on one side. Consider a force balance on a small circular element of radius ρ near the pole when $\Delta\theta$ is small (Figure 16.12). Using the small angle approximation, the radius of this element be $\rho\Delta\theta$, where ρ is the radius of curvature. The stress, σ, on this circular region acts on an area $2\pi\rho\Delta\theta t$ and creates a tangential force equal to $2\pi\sigma\rho\Delta\theta t$. The vertical component of the

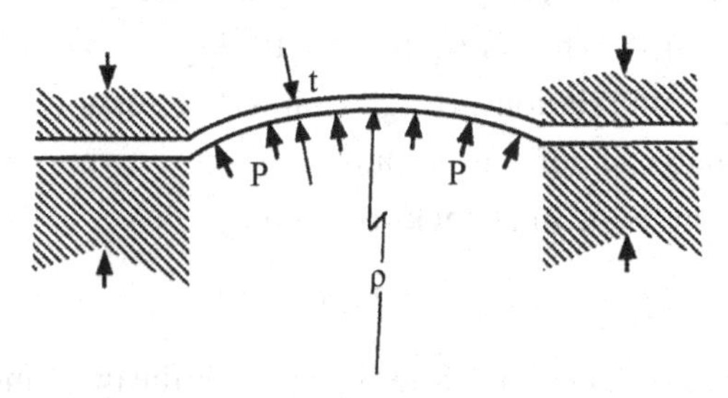

Figure 16.11. Schematic of a hydraulic bulge test of a sheet specimen. Hydraulic pressure causes biaxial stretching of the clamped sheet. From W. F. Hosford, *Mechanical Behavior of Materials*, 2nd ed., Cambridge University Press (2011).

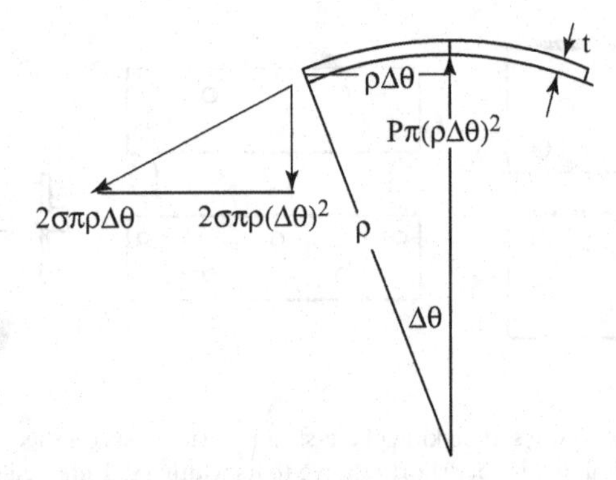

Figure 16.12. Force balance on a small circular region near the dome. The force acting upward is the pressure times the circular area. The total tangential force equals the thickness times the tangential stress. The downward force is the vertical component of this. From W. F. Hosford, *Mechanical Behavior of Materials*, 2nd ed., Cambridge University Press (2011).

tangential force is $2\pi\sigma\rho\Delta\theta t$ times $\Delta\theta$, or $2\pi\sigma\rho(\Delta\theta)^2 t$. This is balanced by the pressure, P, acting on an area $\pi(\rho\Delta\theta)^2$ and creating an upward force of $P\pi(\rho\Delta\theta)^2$. Equating the vertical forces,

$$\sigma = P\rho/(2t). \qquad 16.10$$

Hydrostatic compression superimposed on the state of biaxial tension at the dome of a bulge is equivalent to a state of through-thickness compression.

TORSION TEST

Very high strains can be reached in torsion. The specimen shape remains constant, so there is no necking instability or barreling and there is no friction. Therefore, torsion testing can be used to study plastic stress-strain relations to high strains. In a torsion test, each element of the material deforms in pure shear as shown in Figure 16.13. The shear strain, γ, in an element is given by

$$\gamma = r\theta/L, \qquad 16.11$$

where r is the radial position of the element, θ is the twist angle, and L is the specimen length. The shear stress, τ, cannot be measured directly

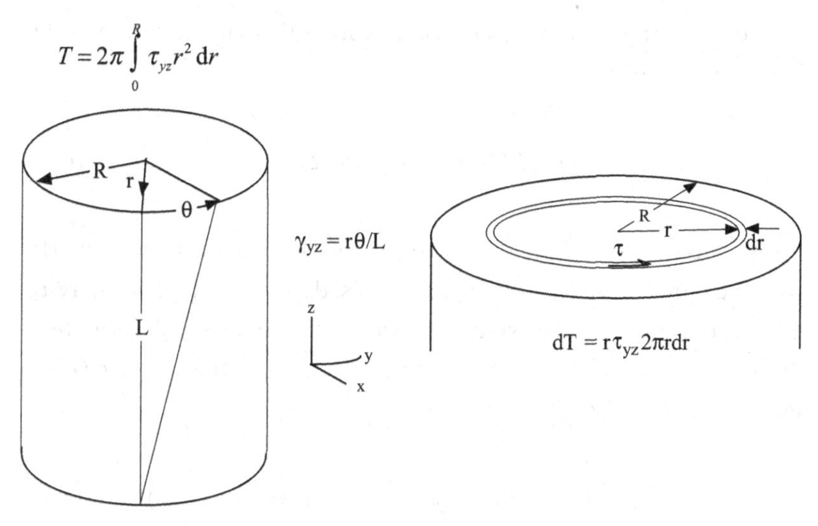

Figure 16.13. Schematic of torsion test. From W. F. Hosford, *Mechanical Behavior of Materials*, 2nd ed., Cambridge University Press (2011).

or even determined unequivocally from the torque. This is because the shear stress, τ depends on γ, which varies with radial position. Therefore, τ depends on r. Consider an annular element of radius r and width dr having an area $2\pi r dr$. The contribution of this element to the total torque, T, is the product of the shear force on it, $\tau \cdot 2\pi r dr$, times the lever arm, r

$$dT = 2\pi \tau r^2 dr$$

and

$$T = 2\pi \int_0^R \tau_{yz} r^2 dr \qquad 16.12$$

Equation 16.11 cannot be integrated directly because τ depends on r. Integration requires substitution of the stress-strain (τ–γ) relation. Handbook equations for torque are usually based on assuming elasticity. In this case, $\tau = G\gamma$. Substituting this and equation 16.11 into equation 16.12,

$$T = 2\pi(\theta/L)G\int_0^R r^3 dr = (\pi/2)(\theta/L)Gr^4. \qquad 16.14$$

Because $\tau_{yz} = G\gamma_{yz}$ and $\gamma_{yz} = r\theta/L$, the shear stress varies linearly with the radial position and can be expressed as $\tau_{yz} = \tau_s(r/R)$, where τ_s is the shear stress at the surface. The value of τ_s for elastic deformation can be found from the measured torque by substituting $\tau_{yz} = \tau_s(r/R)$ into equation 16.12,

$$T = 2\pi\int_o^R \tau_s r^2 dr = (\pi/2)\tau_s R^3, \text{ or } \tau_s = 2T/(\pi R^3). \qquad 16.15$$

Hooke's law cannot be assumed unless all of the deformation is elastic. The other extreme is when the entire bar is plastic and the material does not work-harden. In this case, τ is a constant.

If the torsion test is being used to determine the stress-strain relationship, the form the stress-strain relationship cannot be assumed so one doesn't know how the stress varies with radial position. One way around this problem might be to test thin-wall tube in which the variation of stress and strain across the wall would be small enough that the variation of τ with r could be neglected. In this case, the integral (equation 16.13) could be approximated as

$$T = 2\pi r^2 \Delta r\tau, \qquad 16.16$$

where Δr is the wall thickness. However, thin-wall tubes tend to buckle and collapse when subjected to torsion. The buckling problem can be circumvented by making separate torsion tests on two bars of slightly different diameter. The difference between the two curves is the torque-twist curve for cylinder whose wall thickness is half of the diameter difference.

The advantage of torsion tests is that very high strains can be reached, even at elevated temperatures. Because of this torsion tests have been used to simulate the deformation in metal during hot rolling so that the effects of simultaneous hot deformation and recrystallization can be studied. It should be realized that in a torsion test, the material rotates relative to the principal stress axes. Because of this, the strain path in the material is constantly changing.

NOTE OF INTEREST

Percy Williams Bridgman was born April 21, 1882 and died August 20, 1961. He studied physics at Harvard through to his PhD. From 1910, until his retirement, he taught at Harvard, becoming a full professor in 1919. In 1905, he began investigating the properties of matter at high pressure. After his apparatus broke down, he modified it with the result that he could attain pressures of over 100,000 atmospheres. This led to many new findings, including a study of the compressibility, electric and thermal conductivity, tensile strength and viscosity of more than 100 materials. Bridgman made many improvements to his high pressure apparatus over the years, and unsuccessfully attempted the synthesis of diamond many times. In 1946, he was awarded the Nobel Prize in Physics for his high pressure work.

John Duncan earned a BS in Mechanical Engineering from the University of Melbourne in 1956 and PhD under Professor Johnson from the Manchester Institute of Science and Technology in 1970. He taught at McMaster University from 1970 to 1986, then he became department head at Auckland University. Together with Marciniak, he published a pioneering text, *Mechanics of Sheet Metal Forming*. He developed the details of bulge testing and published. In 1998, he joined a program at Deakins University in Australia and has been teaching at several universities in China.

REFERENCES

1. P. W. Bridgman, *Trans. ASM* v. 32 (1944).
2. J. L. [Duncan, *The hydrostatic bulge test as laboratory experiment, Bulletin of Mechanical Engineering Education*, v.4 Pergamon Press (1965).
3. R. F. Young, J. E. Bird, and J. L. Duncan, "An Automated Hydraulic Bulge Tester," *J. Applied Metalworking* v. 2 (1981).
4. J. L. Duncan and J. E. Bird, *Sheet Metal Industries* (1978).

INDEX

Printed in the United States
by Baker & Taylor Publisher Services